Co-managing complex social-ecological systems in Tanzania

The case of Lake Jipe wetland

Co-managing complex social-ecological systems in Tanzania

The case of Lake Jipe wetland

Christopher P.I. Mahonge

Wageningen Academic
Publishers

ISBN: 978-90-8686-151-4
e-ISBN: 978-90-8686-707-3
DOI: 10.3920/978-90-8686-707-3

First published, 2010

© Wageningen Academic Publishers
The Netherlands, 2010

Preface

Wetlands are important areas in Tanzania. They contain a diversity of natural resources, diversity of fauna and flora, and invaluable habitats for these life forms. Although, in the past, these areas were considered wastelands, the increase in population, and decline in production of terrestrial areas, due to among other factors drought, have led people to migrate to these areas and encroachment and conversion of wetlands to economic activities to optimize fertile and moist soils for agricultural production, and water and pastoral resources for livestock keeping, among others. However, although the wetlands support the people, i.e. by enabling food production, providing pastures for livestock, water for agriculture, livestock and domestic consumption, fishing etc., these uses, if not appropriately conducted, threaten the sustainability of the biodiversity and realization of long-term socio-economic benefits.

In Tanzania, the degradation of natural resources in wetland areas has led to efforts aimed at sustainable use and management of the wetland resources. The preliminary efforts, however, were monopolised by the government, which through its command and control management approaches imposed sanctions to enforce the local community to comply with its top-down policies. Problems were witnessed in the implementation of these initiatives. For instance, due to the lack of a central wetland policy, different sectors intervened at Lake Jipe with different and often contradictory sectorally-based objectives and interests. While one sector may emphasize the conservation of a particular resource, another sector may emphasize its use without integrating environmental concerns. This has exacerbated environmental deterioration instead of curbing it. Although this sectoral perspective persists, in the recent years, the government has recognised that the rural livelihoods and wetland resources are inseparable, and therefore has gradually integrated the local people in the management processes. This shift has been contributed by, among others things, inadequacy in terms of financial and human resources to centralize the control of the vast and diverse wetland resources areas.

Formal natural resources management policies now advocate the integration of local people in the management of natural resources. Nevertheless, there is an endemic lack of knowledge on how to put society-wetland collaboration into practice or how to measure the success of such collaborations in terms of sustainable management. This is particularly the case for the wetland at the centre of this thesis - Lake Jipe. This study aimed to investigate the possibility for co-management between the government and the wetland dependent people for sustainable management of Lake Jipe. The findings from this study may enable us to propose improvements to government-community collaborations for sustainable management of this lake, as well as wetlands further afield in Tanzania and East Africa.

This thesis would not have been possible without assistance, advice, constructive criticism and collaborations from various individuals and organizations. I would like to express my gratitude to The Netherlands Academy of Arts and sciences (KNAW), and the Wageningen University for financially enabling this study.

I extend my sincere thanks to my Promoter Prof. Dr. Ir. Arthur P.J. Mol, and my Co-Promoter Dr. S. Bush, both of Environmental Policy Group (ENP), for their invaluable advise, guidance, encouragement, tolerance, and constructive criticisms from the start to the accomplishment of this thesis. I really appreciate their work and value their role beyond expression.

I also would like to thank my supervisor in Tanzania Prof. P.T.K. Munishi of the Faculty of Forestry and Nature Conservation of Sokoine University of Agriculture. He assisted me with the formulation of original idea on wetlands management, and has likewise assisted me with advice on data collection during the field research in Tanzania.

I extend my thanks to other Staff of Environmental Policy Group for their assistance at various stages and levels. Prof. Dr. Ir. Spaargaren and Dr. Ir. Peter Oosterveer for providing me with insights into social theory, and Dr. Ir. Jan P.M. van Tatenhove for his two courses on Multi Level Governance and Policy Evaluation. They were ready to clarify various issues pertaining to the mentioned subjects at any time, during and after the courses. Thanks to Corry Rothuizen for her administrative and logistic arrangements and support during the whole time I spent at Wageningen University. Really, I appreciate her readily assistance, guidance, and support through the entire time I was in Wageningen.

I cannot conclude my thanks without mentioning the following. I appreciate the assistance of Dorien Korbee for translating the summary of my thesis into Dutch. My office roommates: Elizabert Sargant, Lenny Putman, Jorrit Nijhuis, Hilde Toonen and Dorien Korbee, I thank them for their company, cooperation, and directions at various stages of the PhD research journey. Other friends namely Jingyi Han, Judith van Leeuwen, Michiel de Krom, Hoi van Pham and Dries Hegger, I have enjoyed their company especially during the PhD dinners.

I appreciate the support and encouragement of my wife Damarice, and my daughter Victoria during the research period and when away from Tanzania, for their understanding. Likewise, I should thank my parents, brothers, and sister for supporting my family while I was away.

Table of contents

Map of Tanzania

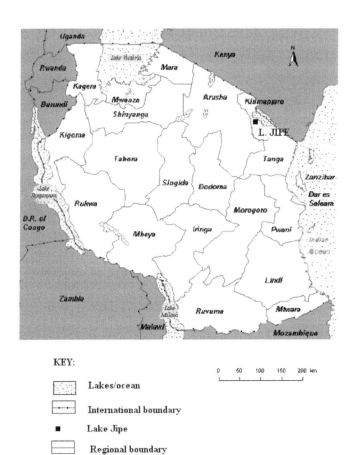

Chapter 1.
Introduction

1.1 Wetland management in Tanzania

Wetlands can be defined as areas of marsh, fen, peat land or water, whether natural and/or artificial, permanent or temporary, with water that is static or flowing, fresh, brackish or salt, including areas of marine water, the depth of which at low tide does not exceed 6 meters (Matthews, 1993). In Tanzania, wetlands cover over seven percent of the country's surface area (Masudi, Mashauri, Mayo, & Mbwette, 2001). Based on the Ramsar Convention definition, Tanzania wetlands are classified into three categories according to their origin and land physiography: coastal wetlands, rift valley systems wetlands, and wetlands of the highland drainage basins. The major wetlands in the country include Rufiji river basin, Ruvu river basin, Wami river basin, Sigi river basin, Umba river basin, Pangani river basin (Lake Jipe integrated), Msangasi river basin, Lake Victoria basin, Malagalasi muyovosi basin, Kilombero, Ruaha, Usangu plains and Ruvuma and Southern river basin.

Wetlands in Tanzania integrate a diversity of biological and socio-economic interests. Biological interests entail provision of support to fauna and flora, some of which are endemic to this wetland area (Twong'o & Sikoyo, 2001). Socio-economic interests include the use of different natural resources for household consumption, irrigation, hydropower generation, fishing, agriculture, animal husbandry, etc. (Bootsma & Hecky, 1993). Due to prolonged drought in some areas, the wetlands in Tanzania have become areas where people obtain their livelihood during these critical times. These livelihood practices include agriculture, livestock grazing, and fishing (Masija, 2000; Salum, 2007). However, inappropriate exploitation (FAO, 1998; Masija, 2000; MNRT, 2003), uncoordinated institutions (Harril, 2002), and pollution (Kassenga, 1997; Shemdoe & Mwanyoka, 2006) contribute to the degradation of the country's wetlands.

Unsustainable human activities carried out at the catchments areas of the wetlands endanger the sustainability of wetlands in Tanzania. Such activities include livestock keeping, household waste disposal, sand mining, fertilizer and agro-chemical application. These activities generate waste and silt that is deposited into water catchments due to ineffective control and disposal institutions and mechanisms (Kassenga, 1997; Shemdoe & Mwanyoka, 2006). When the waste and silts are deposited in wetlands such as lakes, they may affect the water quality in these systems. These changes may favour development of undesired vegetations. Moreover, they may block water streams resulting in decreased water volumes and flows in the downstream areas of wetlands (Shemdoe & Mwanyoka, 2006).

Since wetlands in Tanzania comprise multiple natural resources that are interrelated, disturbance of one natural resource may results in negative effects on other resources. Some parts of wetlands in Tanzania lack certain services at particular seasons due to unsustainable exploitation of resources for other services. For example, water scarcity occurs during the dry seasons, because part of the wetland uses large amounts of water for agriculture thereby degrading water catchment areas (Shemdoe & Mwanyoka, 2006). Consequently, these unsustainable exploitations lead to the deterioration of both ecological and socio-economic potentials wetlands can offer (Harrill, 2002).

In addition to unsustainable and conflicting human practices, poorly designed, implemented and coordinated institutional arrangements (such as policies, legislation, partnerships, codes etc.) endanger the sustainability of wetlands in Tanzania. For example, whereas some natural resources such as mangrove vegetation along the waters of the wetlands are governed by natural resources protection authorities and institutions, industry and trade authorities have the mandate to issue licences for the economic use of wetlands. This results in exploitation of natural resources areas for non-conservation activities (Harril, 2002). Besides contributing to the degradation of natural resources in the wetlands, poor (sectoral) policies and legislation result in conflicts between and among various sectors (Salum, 2007).

Some efforts have been implemented to reduce environmental degradation for the wetlands in Tanzania. For example, the National Environment Management Council (NEMC) of Tanzania started the national wetland conservation and management programme in Tanzania in 1990, in collaboration with World Wildlife Fund (WWF) and International Union for Conservation of Nature (IUCN). These efforts gained momentum during the wetland conservation conference by Southern African Development Cooperation (SADC) in 1991, which urged each member state to formulate its wetland management programme (FAO, 1998). Among the important aims of these programmes, is the integration of the local community in the management of wetland resources. It is believed that such integration may improve sustainable use and management of wetland resources (Salum, 2007). Although international institutions have assisted in increasing the national capacity for wetland protection, national laws and policies have not been formulated (Kassenga, 1997), and the rationale for the government's failure to formulate these laws and policies is unclear (ibid.).

1.2 Managing Lake Jipe wetland

Lake Jipe is a typical Tanzanian wetland. It is not only important for biological diversity but also for socio-economic development. It is a biodiversity rich ecosystem with water-birds including Lesser Jacana, Purple Gallimule, Squacco Heron, Black Heron, African Darter, African Skimmer and fish species namely endemic Tilapia, *Oreochromis jipe* and sardine, *Rastrineobola argentea*. Lake Jipe is also a habitat for crocodile and hippos (MNRT, 2004; Twong'o & Sikoyo, 2001). Socio-economically, the wetland provides livelihood support to the people of Tanzania and Kenya (IUCN, 2000). On the Tanzanian side, more than 120,000 people live around Lake Jipe (MNRT, 2004). Diversity of interests exist among local users including water for irrigation farming, domestic consumption, livestock use, and fishing. Further, the lake is one of sub-catchments of Nyumba ya Mungu dam and Pangani river which generate hydroelectric power.

Since the 1970s the environmental status of Lake Jipe has been under scrutiny. This attention increased from the 1990s onwards when the environmental status of Lake Jipe seemed to deteriorate more seriously and rapidly. Two serious issues regarding environmental problems at Lake Jipe that attracted much national and international environmental attention are the rapid expansion of the waterweeds and therefore increased reduction of the surface area of the lake (IUCN, 2000; TANESCO, 2000), and the drying of the lake which was acute in 2005 (MNRT, 2004). Although much attention was directed to environmental problems of Lake Jipe in the 1990s, degradation of Lake Jipe started in the 1970s when the local people began to experience decreased fishery resources as waterweeds extended over a wide surface area of the lake. The deterioration of Lake Jipe is mainly attributed to the increase of waterweeds and hence the reduction of the surface area of the lake by the community stakeholders. But unsustainable fishing methods, intensive crops farming and livestock management practices have also significantly contributed to the deterioration of natural resources and Lake Jipe wetland as a whole.

Until recently, strategies to address and combat environmental problems at Lake Jipe wetland have hardly focused on human activities in this wetland. In the past years, some research was done on how to eradicate waterweeds using chemical procedures (Gaudet, 1975; Lyatuu, 1981), though to-date such research initiatives have not produced solutions to the waterweeds problems due to among others the fear that chemical application could be hazardous to the fish, humans and wildlife that depend on this lake. This research largely ignored the interactions between economic activities and ecological functions. In other words, humans around the lake were viewed as external entities. This was even more true when involving inhabitants of the wetlands in formulating and implementing solutions to wetland degradation. It can be concluded that local people were viewed as not being able to make a significant contribution to the definition and implementation of solutions for Lake Jipe's degradation. Technical strategies and mechanisms introduced by governmental authorities were regarded appropriate for providing solutions to these resource degradation problems. More recently, this has proven to be impractical and the inclusion of the local resource users in defining solutions and real conservation and management work is now considered imperative (MNRT, 2004).

As a result, strategies which integrate local communities and inhabitants in problem solving strategies came into being. The government and its agencies now consider participation of local communities (natural resource users) vital in the management of natural resources. In 2004, the government of Tanzania through the Tanzania Ramsar Administrative Authority, and with financial support from the International Union for Conservation of Nature (IUCN), initiated an awareness raising strategy (2005-2007) for inhabitants around the wetland, and of additional relevant stakeholders outside the lake region. The aim of the strategy was to elaborate and sensitize stakeholders on their roles, rights and responsibilities in conservation and sustainable use of Lake Jipe resources (MNRT, 2004). This shift from centralized natural resources management to more collaborative and integrated natural resources management assumes that when users are integrated in the management process there is a greater likelihood of not only achieving solution to the present environmental problems but also of having sustainable management process. But sustainability of the management of natural resources and wetland cannot be guaranteed and ascertained by just integrating community actors in management initiatives. Tradeoffs between conservation and social-economic interests, poor capacity in terms of necessary resources and skills of communities, and the institutional and organizational shortcomings of governmental authorities and communities are among the often identified reasons for natural resources degradation at Lake Jipe, even under a more participatory wetland management.

1.3 Studying wetlands as social-ecological systems

This thesis investigates possibilities for co-management between the multiple governmental and community entities for sustainable management of Lake Jipe wetland in Tanzania. In this study wetlands are seen as social-ecological systems. The term social-ecological system (SES) is used to connote a system that views and integrates humans as part of an ecosystem, that is, humans-in-nature. The integration of social and ecological systems is based on an argument that these two systems are interdependent and hence cannot be treated independently if sustainable management of the ecosystem is to be attained (Berkes & Folke, 1998). Before the evolution of this view, humans were regarded as being external entities to ecosystems (e.g. Likens, 1992; Pomeroy, 1988). These views ignored linkages, which exist in practice between humans and natural systems. These linkages are imperative for understanding positive and/or negative impacts on ecosystems due to human and nature interactions (Berkes & Folke, 1998).

Social-ecological systems are made of complex and adaptive social and ecological elements in which ecological components provide feedback and social actors respond to these feedbacks, in order to make the social-ecological systems provide social and environmental interests sustainably (Olsson & Folke, 2004). The approach I take in studying co-management possibilities of social-ecological systems deviates on two points from the majority of social-ecological system approaches. Studies on co-management of social-ecological systems have focused in general on single natural resources management systems (Pinkerton, 1994; Pinto da Silva, 2004; Pomeroy, 1995; Pomeroy, Sverdrup-Jensen, & Raakjaer-Nielsen, 1995; Singleton, 2000). In practice, social-ecological systems often comprise multiple and different but interrelated natural resources management systems, which not only influence one another but also influence the social-ecological system as a whole. Similarly, in most cases co-management studies have addressed government-community collaboration in a singular fashion, whereby both the government and the community are viewed as monolithic units (Pinkerton, 1994; Pomeroy, 1995; Singleton, 2000).

In this thesis, a social-ecological system is viewed as a system that comprises multiple natural resources systems and multiple governmental and community entities and interactions. Hence, this study aims to include these complexities in social-ecological systems. Lake Jipe social-ecological system therefore is viewed as comprising multiple interactions of multiple social and natural units and systems. Therefore, I consider it vital to include interactions between downstream to upstream geographic areas and other spatial linkages, to include different scales from local to international, and to include multiple natural resource disturbances, multiple actors within both government and community and multiple institutions for managing these disturbances and conflicting interests. Only through including these linkages I will be able to understand problems with sustainable management of Lake Jipe wetland. For example, inappropriate farming practices on the upstream areas may pollute water and degrade fisheries resources on the downstream. Analyses that are confined within sectoral boundaries of one natural resource run the risk of misunderstanding linkages that affect sustainability of the natural resources systems individually and the social-ecological system as a whole.

Including cross-sectoral, cross-scale and actor diversities that exist in social-ecological systems also contributes to new theoretical schemes and knowledge on natural resources co-management. It will also have consequences for the recommendations on mechanisms and strategies that can be promoted and/or adopted for sustainable management of the multiple natural resources systems and the social-ecological system as a whole. In that sense, Lake Jipe wetland is a case study for developing new approaches to study complex social-ecological systems. In the following section, the thesis objectives and research questions will be given.

1.4 Thesis objectives and research questions

At Lake Jipe, and also more widely in Tanzania, there appears to be a change from simplistic state centred natural resources management approaches to natural resources governance regimes in which multiple governmental agencies and community entities collaborate and interact in the management process. As these governance approaches evolve, there is a need to understand their roles and effects on the sustainable management of Lake Jipe social-ecological system. There has not been any study carried out on how interactions and collaborations between the diversity of social actors (both within the community and the government) and the various ecological resources (fisheries, land and water) impact on and affect natural resources management of the Lake Jipe. This thesis aims to fill this gap. In doing so this study challenges and extends the existing theoretical knowledge on natural resources governance, as well as providing recommendations for the further improvement of wetland (co-)management at Lake Jipe and more widely in Tanzania.

The central aim of this research is, therefore, to investigate co-management arrangements on multiple natural resources, involving multiple governmental and community entities for sustainable management of Lake Jipe social-ecological system. In order to achieve this central objective, this research addresses the following research questions:

1. How do government and community collaborate and interact in managing natural resources in livestock production, agriculture, and fisheries, and how do their interactions influence natural resources management?
2. What co-management arrangements emerge from interactions between governmental and community institutions and actors around livestock, agriculture and fisheries?
3. How do the different co-management arrangements relate, interact and influence one another, and what are the implications of their relationships and interactions on managing Lake Jipe social-ecological system as a whole?

1.5 Structure of the thesis

The thesis is divided into six chapters.

Chapter two introduces the theoretical framework around which the thesis is based. Theories on co-management of natural resources between the government and the local community are used as a starting point to build a conceptual model for analyzing natural resource co-management at Lake Jipe. In this chapter, three concepts – namely *arrangements*, *institutions*, and *actors* – are used for analyzing co-management arrangements at Jipe social-ecological system. In addition, the institutional dimensions of co-management arrangements – *empowerment, conflict, scale, participation, heterogeneity, property rights, and leadership* – are reviewed. Following this chapter, an introduction on lake Jipe (the study area) including a methodology and data collection is given.

Chapter three provides an historical overview of the institutional changes in natural resources management in Tanzania. Different eras of natural resources use and management, and of interactions between governmental and community institutions and actors, are examined. These eras entail pre-colonial, colonial and post-colonial periods. This chapter also introduces the current government natural resources administrative structure in Tanzania.

Chapters four, five and six present the three empirical cases of this thesis. Chapter four analyzes collaborations and interactions between and among the governmental and community institutions and actors in managing and mediating conflicts in fisheries management at Lake Jipe. Interactions among and between governmental actors and resident and non-residents fishers are analyzed. Consequently, this chapter identifies and evaluates co-management arrangement in fisheries management involving the governmental, community and hybrid institutions and actors.

Chapter five analyzes collaborations and interactions between and among the governmental, non-governmental, and community institutions and actors in the use of natural resources in agriculture. It highlights the collaboration of multiple and diverse institutions and actors in agricultural resource management and conflict mediation. This chapter likewise identifies and evaluates co-management arrangement emerging in the use of natural resources for agriculture.

Chapter six analyzes collaborations and interactions between and among governmental and community institutions and actors in the use of natural resources for livestock production. It highlights interactions between the ethnic institutions and actors and between these actors and the governmental institutions and actors in the use and management of pastoral resources. This chapter ends with identifying and evaluating co-management arrangements between the governmental and community institutions and actors in the management of natural resources in livestock production.

Finally, chapter seven presents the conclusions of this thesis. An answer to the question how governmental and community institutions and actors co-manage natural resources at Lake Jipe social-ecological system is given. First, the summary of findings from the three empirical cases is outlined, and then the three co-management arrangements are compared to unveil similarities and differences, collaborations and conflicts and incompatibilities in the use and management of natural resources at Lake Jipe. Finally, the conclusion is given highlighting the implications of the findings from Lake Jipe to natural resources management in Tanzania as a whole, and its contribution to co-management theory.

Chapter 2.
Co-management of natural resources between the government and the local community

2.1 Introduction

The objective of this thesis is to analyse natural resource management as a joint effort of the government and the local community at Lake Jipe in Tanzania. Specifically, it analyses relationships between the government and the local people in the management of natural resources for agricultural production, livestock production and fisheries management. The aim of this chapter is to develop a theoretical framework for the co-management of natural resources that is instrumental in analysing the interactions between multiple governmental entities and multiple community entities in the management of multiple natural resource systems in one social-ecological context, Lake Jipe. This chapter introduces the complexity of sustainably managing social-ecological systems and the role of co-management arrangements, institutions and actors.

Studies on the co-management of natural resources involving the government and the local community have been conducted in a simplistic fashion such that the government and the local people are viewed as single entities (Pomeroy, 1995; Pomeroy & Berkes, 1997; Pomeroy, 1998; Pomeroy, Katon, & Harkes, 2001; Sen & Nielsen, 1996). In the same way, social-ecological systems have been regarded as one unit. In practice, the government is comprised of multiple agencies and the community of multiple actors; what looks like one system consists of various entities (various social-ecological systems within one system) (Pomeroy, 1995; Pomeroy, 1998; Singleton, 2000). To conceptualise co-management as comprised of a unitary government and a unitary local community is increasingly becoming unrealistic. Co-management under such a simplistic conceptualisation falls short of genuine reflection and fails to address issues stemming from internal politics in the community due to the existence of different and sometimes contrasting interests associated with various user groups. By the same token, because the government is made of different agencies/units (e.g. agricultural, livestock, and fisheries), conceptualising the government as unitary is liable to cause a failure to accommodate contrasting interests and politics as arise in the interactions among these diverse government units. Similarly, if one regards a social-ecological system as one unit, one is likely to misunderstand the diversities and complexities in social-ecological systems.

In this thesis, I am going to unpack the co-management arrangement into three parts whereby the government and local community relationships in the management of lake Jipe wetlands are studied in terms of agricultural production, livestock production and fisheries management.

This chapter is organised as follows. In section 2.2, an overview of the evolution of natural resources management is given. This is based on the rationale that co-management arrangements that exist today are a result of developments that have occurred in the field of natural resources management. It is, therefore, worthwhile to cultivate an evolutionary picture of natural resource management. In the discussion on the evolution of natural resources management, three approaches are discussed: classical, neo-liberal and populist approaches. Because this thesis addresses the co-management of natural resources between the government and the local people – in other words, relationships between the two levels of management of natural resources – the focus is on the relationships between the government and the local people. Specifically, in these approaches, I look at how the two levels have participated in resource management and how the government relates the local community to its natural resources (environment) and vice versa. I also review the strengths and weaknesses surrounding these approaches to the management of natural resources. Section 2.3 introduces the rationale for co-management arrangements involving the governmental and community levels. In this section, conflict as one of main issues surrounding resource co-management is introduced. Section 2.4 defines analytical concepts, namely arrangements, institutions, actors and co-management relevant for analysing the collaborative management of natural resources in the Lake Jipe social-ecological environment. In section 2.5, I use the following institutional dimensions – *empowerment, conflict, scale, participation, heterogeneity, property rights, and leadership* – to analyse the role of institutions in governing the relevant actors in the co-management of natural resources. This section concludes by introducing the complexity of the lake Jipe social ecological system whereby one system demands three co-management arrangements.

2.2 Evolution of natural resources management

Until colonial times, common pool management systems in traditional communities existed. The management of natural resources was based on clan-oriented arrangements. Rights, rules, conventions, obligations and sanctions accompanied these traditional management systems (Adjewodah & Beier, 2004; Opoku-Ankomah, Ampomah, & Somé, 2006). In Africa, for example, natural resources such as land were held under the tutelage and stewardship of chiefs and clan leaders (Gwebu, 2001). The coming of colonialism marked the onset of the replacement of the traditional management systems (Guha, 1997). This section will review three sets of theories about how natural resource management has evolved since colonial times. These are the classic approaches, neo-liberal approaches, and populist approaches. The neo-liberal and populist approaches emerged parallel to each other.

2.2.1 Classical approach

The classic approach dominated from the 1950s to the 1970s. It is a top-down approach related to environmental management and rural development. It was developed through government-sponsored scientific institutions and was applied through extension agents (Biot, Blaikie, Jackson, & Palmer-Jones, 1995; Blaikie 1996; Blaikie, Brown, Stocking, Tang, Dixon, & Sillitoe, 1997). The central thesis of this approach is that the use of common property by the local people results in the tragedy of commons (Hardin, 1968). In the following paragraphs, this approach is analysed to consider how local people participated in natural resource management under its dominion and what the relationships between the government and the local people were like. In addition, critiques of the classic approach will be revealed.

The classic approach views the local people as a threat to the environment. Their institutions, traditions, rules and norms are seen as destructive to the environment because they are non-scientific, superstitious and illogical, and they should therefore be replaced by formal institutions.

At odds with conservation objectives, local people aim at maximising individual gains at the expense of the environment. They should be instructed by the government through technological and policy interventions whereby encouragement, persuasive and coercive means are employed to compel them to comply with the interventions (Biot *et al.*, 1995; Blaikie, 1996; Blaikie *et al.*, 1997). In other words, from the classic perspective, the view exists that sustainable natural resource management will be achieved through the nationalisation and privatisation of the resource, not via community-based management regimes.

The classic approach, furthermore, reduces local people to environmentally irrational beings. Although these individuals degrade the environment, they do not identify its problems. Those who identify the environmental problems are instead the external agents (government, donors and researchers), and the solutions come from these external agents, whereas the local people are required to cooperate by implementing solutions. In other words, local knowledge is not recognised. Mechanisms for ensuring the cooperation of the local people range from encouragement to persuasion and the use of threats (Pelosikoti, 2003). The participation of local people in natural resource management therefore, according to the classic view, must be confined to implementing whatever programs the government experts tell them to.

Several critiques exist of the classic approach. Ostrom and colleagues (1999) assert that although the tragedy of commons undoubtedly occurs in some areas, generalisations should be avoided. There are cases wherein local people co-exist with natural resources for thousands of years and develop sustainable institutions for governing the commons. These authors have reported empirical cases wherein grazing lands under the intervention of the state and private management regimes (e.g. in China and Russia) were more degraded than grazing management regimes under a group of pastoralists (e.g. in Mongolia).

Ostrom (1990) posits that individuals engaged in collective action have their own mechanisms and institutional arrangements that have to be considered when a top-down model is imposed on a given resource area. The local institutional arrangements, she argues, are instrumental to the sound management of the resources, and a failure of collective management options occurs when an externally imposed intervention fails to consider these institutions. According to Ostrom, theoretically devised models cannot be successful in practice (in the 'real world') if they are incompatible with local institutional arrangements. In other words, Ostrom implies that new institutional arrangements must build on the existing institutional arrangements to be successful. However, building on and providing support to local institutions is not always a guarantee for attaining resource management objectives. There are cases where, despite the decentralisation of decision-making from the government to the local communities and institutions, the expected resource sustainability objectives have not been attained (Agrawal, 2001). This has been due to the existence of internal politics in a largely heterogeneous community (Le, 2004; Leach, Mearns, & Scoones, 1999). Other authors criticise Ostrom's view of local institutions as being focused more on the internal aspects of the community and as not looking at external factors that may compel the community to act individually at the expense of others. Such situations include extended crises that may make the individuals in question adopt selfish strategies to preserve future income (Baland & Platteau, 1996).

Sekhar (1999) asserts that the generalisations by proponents of the classic approach that local institutions degrade natural resources are not always true. Citing empirical cases from India, he claims that traditional institutions regulate resource use and co-exist with natural resources. Sekhar found that traditional institutions attach resource use to social authority and beliefs, about which it is believed that a lack of compliance results in misfortune. Sekhar also asserts that informal institutions that control resource use become more powerful based on the scarcity of a particular species in an ecosystem. For example, the author tells how some valuable scarce tree species (*Dalbergia sisoo* and *Dendrocalamus strictus*) are strictly protected by the imposition of higher fines by the local people. This case indicates an integration of informal institutions (restrictions) and market incentives (fines) to avoid the over-exploitation of endangered resources. However, the author reports that when the government intervenes, the resources become degraded. This, he claims, is because the power of traditional institutions is undermined by government intervention.

Another criticism has to do with the assertion that top-down government intervention is the solution to the unsustainable management of natural resources. Ostrom (1990) asserts that there are higher costs incurred by the government in enforcing compliance with sustainable natural resource practices than noted in the simplistic generalisations of the classic approach. In other words, the government does not have adequate resources (financial or human) to adequately supervise and enforce top-down natural resource management regimes. However, despite this higher transaction cost, some scholars (e.g. Uphoff, 1998) have shown that successful local management systems are usually not operating in isolation from other governmental and non-governmental institutions and organisations. In other words, the issue of transaction cost does not rule out the limited capacity of local institutions to sustain sound natural resource management alone.

Other scholars (e.g. Feeny, Berkes, McCay, & Acheson, 1990) claim that common property regimes are in fact instrumental in controlling access to and use of resources. They assert that the tragedy of the commons occurs after former communal property rights are undermined and changed to allow open access. Examples of undermining factors, according to these authors, are colonialism, overpopulation, changes in technology, and changes in the economy, such as new market pressures.

Along with the above critiques, in practice, there have been conflicts between the government and the local people over natural resource management. For example, Girot, Weitzner, and Borrás (1998) explain how in Costa Rica, a country of peace as they call it, the top-down management regimes resulted in conflict between the people and the government when local people resisted relocation from the national park. Similarly, experience has indicated that top-down natural resource management policies cannot successfully manage natural resources on their own. External developments, including population growth, changes in consumption patterns, and globalisation, which have affected the culture and markets, have rendered classic approaches insufficient. Under these influences, government policies need input from other stakeholders who may have different interests, have different perceptions, and control different types of resources. The top-down government policies thus have become inadequate and need to accommodate other perspectives and stakeholders in policy-planning processes (Fresco *et al.*, 2005). This has triggered and mobilised the consideration of alternative natural resource management approaches. Neo-liberal and populist approaches have thus emerged.

2.2.2 Neo-liberal approach

The neo-liberal approach, developed by the World Bank (World Bank, 1990-1995) among others, stressed a shift from the state-controlled management of natural resources to giving more power to the market (through deregulation) to create incentives for the conservation of the environment. Such incentives include subsidies, taxes, a market-like pricing system and the creation of private property right systems (Biot *et al.*, 1995). In the following paragraphs, I analyse how the community partakes in natural resource management under a neo-liberal approach and consider the government and community relationships in contexts of economic liberalisation. As before, critiques of this approach will also be discussed.

The neo-liberal approach holds that the participation of the local people in biodiversity conservation is regulated by economic incentives such as markets, policies and institutions (Adger, 2001). This includes the pricing of externalities resulting from the unsustainable human use and management of the environment. Local people are viewed as economically rational beings, each driven by a desire to maximise profit from the use of natural resources (Post & Snel, 2003). Local institutions are important insofar as they provide information that is used as a basis for making decisions regarding natural resource use and management. The market logic creates economic incentives such as taxes, subsidies, and prices. When these instruments fail, regulations are imposed to govern the sustainable exploitation of natural resources (*ibid.*).

Following the development of this approach, contrasting empirical evidence has been reported. On the one hand, some studies reported a constructive and important role for the market in resource management, which might, for example, formalise environmentally friendly traditional institutions or invent new ones where they did not previously exist (Gemma, 2001). Other scholars (Nasi *et al.*, 2008) have indicated that in some countries, such as Malaysia, degraded resources are being restored through the use of regulations, publicity and education programs, the control of external encroachment, allowing subsistence use of natural resources and monitoring markets, shops and restaurants to control illegal unsustainable exploitation. This has indicated that in some cases, the market approach not only can control unsustainable exploitation but also can sustain the subsistence use of resources on the part of local users.

In contrast, other scholars (e.g. P. Ehrlich & A. Ehrlich, 1991) have reported negative influences of the market on natural resources. According to them, the advancement and increase of market channels goes hand-in-hand with the advancement of roads and transport networks linking and integrating the local community into larger market systems, which results in resource and environmental degradation (Chomitz, 1995; Nilsson & Segnestam, 2001; Young, 1994). They assert that as market forces become firmly established, local users are motivated to harvest much greater quantities of the resources because they are now exploiting resources for cash (Nilsson & Segnestam, 2001; Stocks, 1987) and that therefore, the combined effects due to subsistence and market forces hasten environmental degradation. Moreover, the market forces create and maintain inequalities among people whereby rich people benefit from the use of natural resources more than do poor people. This marginalises the poor, with further resource degradation as a result (D. Brown, 2003). The poor may become displaced from productive resources (e.g. land), and thus move to marginal fragile areas, consequently extending and exacerbating environmental degradation (Barbier, 1997).

Conflicts have also been experienced in the implementation of the neo-liberal approach. Some researchers have reported the emanation of conflicts in the privatisation of natural resources or the communally or publicly owned environmental goods and services. Frakin (2001) claims that the formalisation of private property rights interferes with communal resource tenure arrangements. Such formalisation, he asserts, may create classes within the community because some individuals may successfully acquire a property and convert it to private investments while others may become propertyless. Dismantling the communal ownership arrangements transforms a social setting into an arena of conflict between the haves and have-nots (Kanyongolo, 2005). Besides, conflicts may occur not only among resource users but also between the government and resource users. Such can be the case when the privatisation of a natural resource results in restricted access to some environmental goods and services by local users (Spronk & Webber, 2007).

Several criticisms of the neo-liberal approach exist. First, it reduces local people to economically rational entities (Pelosikoti, 2003). The neo-liberal approach sees market-based incentives as ensuring the sustainable management of natural resources. That is to say, when there is a good market for agricultural produce, for example, the local people will be motivated to adopt sustainable production technologies. In other words, local people are always striving to attain maximum economic gains at the expense of the environment. This approach highlights that there are already sustainable technological options but that a clear understanding is required of disincentives to their adoption as seen by the local people. Additionally, this approach is criticised for its unclear definition of the best technological options (Blaikie *et al.*, 1997). Second, the neo-liberal approach is criticised for the irrelevance of its economic incentives to the local people. The criticism is built on two aspects. On the one hand, there is a mismatch in benefits between the local people and conservationists despite the fact that the former have had the primary interaction with natural resources for many years. On the other hand, the neo-liberal approaches are reinventing top-down approaches to natural resources management and are not really constructed upon norms of democratic participation (K. Brown, 2003).

2.2.3 Populist approaches

Populist approaches emerged in the 1980s in parallel with market approaches and following the failure of top-down natural resource management regimes. It integrates development and conservation objectives to attain sustainable development (Kumar, 2005). This concept evolved following the Brundtland Commission (WCED, 1987) and the United Nations Conference on Environment and Development (UNCED, 1992), which argued for the integration of nature conservation with human needs. In other words, the approach advocates the active involvement and empowerment of local communities as a pre-requisite for sustainable development and natural resource management. The emergence of this approach was influenced by the existence of empirical evidence indicating that traditional resource management systems in common property regimes have contributed to protecting natural resources from being over-exploited and have played a significant role in biodiversity conservation (Bromley, 1992; Ostrom, 1990).

Contrary to classic and neo-liberal approaches, the populist approach advocates for bottom-up natural resources management and values local institutions as an important element of sustainable natural resource management. These approaches regard local people as rational beings not only from an economic point of view but also from an environmental perspective. They advocate bottom-up participatory planning for natural resource management (Post & Snel, 2003). The emphasis is on the empowerment and participation of the local people as key to sustainable development and conservation (Shukla, 2004). Populist approaches emphasise rights, justice, self-determination and empowerment as crucial for attaining sustainable or long-term environmental management. They see local people as an appropriate focus and level for sustainable environmental conservation efforts (Adger, Brown, & Tompkins, 2005).

However, populist approaches are overly optimistic, providing a simplistic view of the community, and therefore mask the reality of the community. It is assumed that when natural resource initiatives fall under jurisdiction of the local community, their success is guaranteed (Cheong, 2004). The capacity of the local people to use natural resources on a sustainable basis is exaggerated (Spinage, 1998). There is also an over-simplified idea that the degradation of the environment by local industry is a result of the marginalisation of local people by powerful political, social and economic forces outside the community (Adger, 1999). Evidence nonetheless exists on initiatives that have failed despite their being community-based (Agarwal, 2001; Fajber, 2005). This has led to thinking among theorists regarding the reasons for the failure of this widely promoted and accepted approach.

Empirical research has uncovered that the community is made up of groups of heterogeneous people with different interests and goals. This diversity has an impact on decision-making regarding natural resource management (Agarwal, 2001; Cornwall, 2000; Fajber, 2005). Such relations and heterogeneity among social groups are never static but instead keep on changing on both spatial and temporal dimensions, and they do not end at the community level but instead scale down to an individual household level (Cornwall, 2000). Cheong (2004) asserts that in the community, there are diverse groups with diverse interests and dynamic internal politics such that sometimes, especially in times of crisis, the community needs external support and resources. Thus, they have to create linkages with external institutions and actors. Biot and colleagues (1995) further argue that heterogeneity exists even in the knowledge possessed by social groups and individuals based on gender, occupation, age, social status and class. The result of this heterogeneity is conflict, which requires attention in natural resource management.

The populist approaches have also failed to reflect the diversity of actors with cross-scale relationships. Although local people are heterogeneous in various respects, such as with regard to income, goals, interests, ethnicity, social status etc. (and of course these elements determine the power relations among them and the environment), populist advocates ignore these cross-linkages at the level of the community (McNab, 2004; Peet & Watts 1996). Additionally, government-community interaction, which can be useful in mediating negative linkages among resource users, is viewed as an avenue for marginalising the resource users and disrupting their alleged positive relationships with the ecosystem (Belsky, 1999). Populist proponents, therefore, suggest less interference by the government. This is supposed to give the people autonomy by allowing them to apply their local experience, wisdom, and knowledge in their interactions with the environment (Ascher, 1995; Douglas, 1992; Western & Wright, 1994).

2.2.4 Strengths and weakness of classical, neo-liberal and populist approaches

The three approaches have potential and pitfalls. As regards potential, they all have to do with natural resources degradation and therefore are relevant in addressing environmental problems, although they differ in identifying the cause of environmental problems. While the classical approaches view local people and institutions as the source of environmental degradation, the neo-liberal approaches assert that the problem of environmental degradation is the absence of effective economic incentives. On the other side, populist approaches view environmental degradation as resulting from interference in sound local institutions on the part of the government and private institutions.

All three approaches propose ways of addressing these problems, although the proposed solutions differ. The classic approach proposes top-down technological interventions and the nationalisation and privatisation of management initiatives as a solution to natural resource degradation – i.e. the exclusion of local people and institutions from management regimes. On the other hand, the neo-liberal approach considers the participation of the local people in natural resource management through the creation of market incentives. In other words, this approach reduces local people to economically rational actors. The populist approach sees the solution to environmental degradation as giving resource users and institutions a central role in environmental and natural resource management.

The three approaches also have various weaknesses. Classic approaches are limited in terms of human and financial resources, rendering the government incapable of adequately combating environmental degradation and ensuring socio-economic benefits for its people. Neo-liberal approaches, despite the creation of economic incentives for motivating the conservation of natural resources, are viewed as a top-down approach. The incentives advocated in these approaches are also perceived as both inadequate for and also perhaps irrelevant to the community perspective. The populist approach is too quick to assert that the community can manage natural resources on its own in a sustainable way. Research has shown that this is not always the case; some resources have been undermined and degraded despite the programs' being community-centred.

The implementation of the three approaches has entailed different empirical outcomes as I have seen in the review above. While the approaches have positive outcomes in some places, they have negative consequences in other places. This implies that we cannot generalise regarding the success or failure of these approaches for all environmental contexts. In situations in which both state-centred and community-centred management regimes have not yielded the expected outcomes, this has led to the reformulation of approaches that integrate the community and the government in natural resource management. In other words, co-management approaches result from situations in which separate natural resource management regimes under the government (Gehab & Crean, 2000; Hara, Donda, & Njaya, 2002) and the community (Hachongela, Jackson, & Sen, 1998; Kebe, 1998) have not yielded the expected outcomes. One such obstacle is conflict within and between the government and the community. With co-management approaches, it is assumed that the strengths of each institutional actor (the government and the local community) complement the weaknesses of the others (Pomeroy & Berkes, 1997). There is, therefore, a need to revisit natural resource management arrangements, institutions and actors, to understand how new interactions have been created. The next section now addresses issues of co-management and conflict.

2.3 Co-management between the government and local community

In this section, I explain the essence of co-management. A brief definition of co-management according to the co-management literature is given, followed by reasons for the establishment of co-management arrangements involving governmental and community institutions and actors. The critical issue highlighted in this respect is the mediation of conflicts among resource users in natural resource management.

For some, co-management is the sharing of power and responsibilities between the government and the local people (Kumar, 2005; Pomeroy & Berkes, 1997). For others, it may be that the government holds the power (decision-making) but shares management functions, entitlements and responsibilities among users of natural resources in a given area (Borrini-Feyerabend, Farvar, Nguinguiri, & Ndangang, 2000). Yet for other authors (e.g. Nielsen & Vedsmand, 1999; Sen & Nielsen, 1996) co-management is the sharing of responsibilities only between the government and the community. Co-management can also take the form of Pinkerton's horizontal folk-managed systems and vertical contracting out of the state management powers model (Pinkerton, 1994). In this case, there is a horizontal continuum of government and community management systems. On one side of the continuum is a near-total state management system, whereas on the other side of the continuum is near-total self-management system. At the same time, as the vertical model advocates, the government may award rights at the community level (Pinkerton, 1994). In principle, therefore, co-management is the collaboration between the government and the people, where they may share powers, responsibilities, management functions, rights and entitlements.

Government and community natural resource management systems rarely adequately act alone to attain a successful natural resource management initiative. According to Gibbons (1999), the top-down natural resource management system is blunt and insensitive to the opportunities and constraints of local situations. On the other side, Cash *et al.* (2006) assert that the bottom-up approach is insensitive to the contribution of actions of the local people taken to address large-scale environmental problems. These authors agree that neither the government nor the community can resolve natural resource management problems alone. One strategy for resolving these problems, they argue, might be the integration of the two systems – i.e. the government and the community. In other words, solutions can be attained through the co-management of natural resources involving the government and the community arrangements, institutions and actors.

Problems or challenges that exist in government or community natural resource management systems create driving forces for co-management. Adger *et al.* (2005) claim that incentives must be created to facilitate the creation of co-management arrangements. The authors argue that the local community may enter in a co-management arrangement when, for example, there exist historically marginalising power relations among users of natural resources. The community may feel marginalised by external resource users and may anticipate that forming a co-management agreement with the government may help in protecting its interests.

Co-management may also reduce the burden on individual actors. The community may take on roles formerly played by the government (e.g. the monitoring of sustainable resource use, the appropriate use of harvesting gear, etc.). The government may retain the responsibility of creating a favourable environment for enacting the roles entrusted to the community. Such facilitative functions of the state could include creating conducive legislative, administrative and judiciary contexts (Buttel, 1998). In some cases, community instruments for conflict management might be weak, partial, or unable to address and manage situations with high levels of conflict. In these cases, the government may play an important role in fulfilling those functions that the community cannot (Pomeroy *et al.*, 2001). Furthermore, the government may protect wider public goods such as watersheds, biodiversity and carbon sinks, as well as facilitating and regulating private activities (Shackleton *et al.*, 2002).

According to Jentoft (2004), the integration of the government with the community in the management of natural resources is important. He argues that for any natural resource, there are a web of social interactions between the immediate social community around that resource and other stakeholders away from that area. The external stakeholders, in one way or another, depend on the same resource because the resource is important for the immediate community's interests as well as those of the general public. The government should therefore stand for the interests of the general public. This, in turn, may justify the integration of the government in natural resource management arrangements as an alternative to leaving the community on its own to manage a particular natural resource.

The problem in natural resource management may be not just the degradation of natural resources but also the existence of conflicts among resource users. The term conflict may have different meanings depending on the context. In the context of natural resource management, conflict can be defined as a situation wherein two or more social entities or parties have incompatible purposes and interests and, therefore, hostile attitudes emerge or one party takes action that may undermine the ability of another party to address its interests and purposes (Mitchell, 1981).

Conflict may occur because of diverse behaviour, preferences, interests, and objectives among institutional actors (Kumar & Kant, 2007). Conflict may also stem from resource use interactions in which one or more users feel discontent, marginalised or unfairly treated (Christie, Buhat, Garces, & White, 2003a). In other words, heterogeneity in various respects may be the main cause of the emergence of conflict between institutional actors. In other cases, conflict occurs because of a weak or absent government regulatory authority (Isaac, Ruffino, & McGrath, 1998). Similarly, conflict may result from the violation of governing institutional arrangements (Pomeroy *et al.*, 2001). This conflict can cause personal violence and sometimes even armed confrontation among the resources users, especially when the resource involves different user categories delineated by geographic areas: for example, users who reside within a certain resource boundary versus outside users (Pomeroy *et al.*, 2001; Isaac *et al.*, 1998). Conflict may render unsuccessful concerted efforts at sustainable use of natural resource. Individual governmental and community institutions and actors may fail to resolve or mediate these conflict situations. Such conflict is a primary conflict because it occurs before co-management arrangements are devised. Co-management strategies involving the government and the community may be formulated to collaboratively resolve conflict (Hachongela *et al.*, 1998; Jackson, Muriritirwa, Nyakahadzoi, & Sen, 1998; Kebe, 1998; Sowman, Beaumont, Bergh, Baharaj, & Salo 1998).

However, there is a chance of introducing conflicts while implementing co-management arrangements as well. This kind of conflict can be deemed a secondary conflict. Singleton (2000) argues that when co-management arrangements are implemented, conflicts may emerge not only between the government and the community but also within and between the community users. She posits that although co-management in most cases constructs institutional arrangements and mechanisms for resolving conflicts, it might result in strengthening existing conflicts and even creating new ones. The newly established or modified institutional arrangements in co-management, she argues, may marginalise some actors, especially less powerful groups or individuals in the community. Pomeroy *et al.* (2001) assert that when new institutional frameworks are installed, it takes time for them to become adapted to a given community, and that during this transition, conflicts may persist or emerge in the community. Therefore, dependable mechanisms for mediating potential conflicts that are acceptable to the parties involved must be available during this transition.

Conflict, therefore, is one of the issues that may influence the formation of co-management arrangements involving government and community institutions as well as actors to mediate such conflicts and improve sustainable and collaborative natural resource management. In such a situation, various governmental/community and formal/informal institutions and actors partake in the management of natural resources in a social-ecological environment. Because this thesis analyses co-management entailing multiple governmental and community entities in one social-ecological system, some powerful tools – namely, arrangements, institutions and actors – are needed to analyse the interactions among these entities. In the next section, these tools will be defined.

2.4 Concepts for analysing co-management of natural resources

This section defines the concepts that are used in analysing co-management arrangements involving multiple government entities and multiple community entities in the management of multiple natural resource systems as part of the Lake Jipe social-ecological system. These concepts are co-management, arrangements, institutions, and actors (Figure 2.1).

I use the following concepts – *arrangements, institutions* and *actors* – to analyse co-management arrangements in the Lake Jipe social-ecological system. Natural resource management systems are made up of various institutions that govern and mediate the practices of actors that occur within some arrangements, and therefore, an analysis of co-management at Lake Jipe cannot omit them. While these concepts regularly emerge in co-management literature, they are often confused with one another. In this thesis, however, these concepts are used for analysing co-management arrangements, and through this analysis a clear definition of these concepts will be presented.

Co-management in this thesis implies a collaboration involving formal/informal, and government/community institutions in which governmental and non-governmental actors collaborate in managing natural resources and mediating conflicts around natural resource management.

A co-management arrangement implies a sectoral management system (e.g. agriculture, livestock or fisheries). Within the arrangement, multiple governmental/community and formal/informal institutions govern the practices of governmental and community actors in managing natural resources and mediating conflicts that emerge in the process. In some arrangements, non-governmental actors (NGO) may link and build up the capacity of government and community

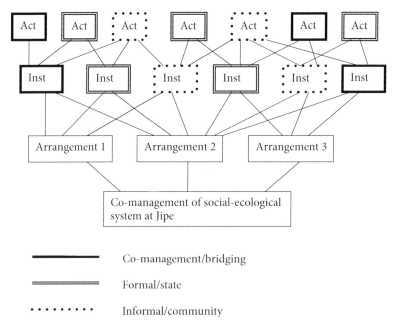

Figure 2.1. Analytical concepts for analysing co-management of natural resources in Jipe social-ecological environment.

actors. These non-governmental actors may have bridging functions horizontally at the community level and vertically between the government and the community institutions to enable them to implement sustainable management decisions and practices.

Institutions can have multiple and sometimes contrasting definitions according to the context in which they are implied. According to North, institutions are a 'set of rules, compliance procedures, and moral and ethical behavioural norms designed to constrain the behaviour of individuals' (North, 1981: pp. 201-202 cited in Feeny, 1988, p. 171). Ostrom (1992) defines institutions as a set of rules specifically used by a set of individuals to organise repetitive or routine activities that produce outcomes that affect these individuals and others. North (1993) also sees institutions as constraints the humans devise to structure or govern their relationships. These constraints are formal (e.g. rules, laws and constitutions) and informal (e.g. norms, conventions, and self-governing codes), and they include implementation and enforcement characteristics. Peters (1999) defines institutions as a collection of values, rules and repetitive actions that are devised to implement and enforce those values.

In this thesis, institutions are defined as rules, norms, conventions, and customs governing and linking the practices and decisions of users and enforcers of natural resource management. The institutions in this context can be formal, informal and hybrid (co-management) institutions. A formal institution implies rules, laws, and regulations devised and imposed by the government (national, regional, and/or local) for governing the management of natural resources at the community level. On the other hand, an informal institution implies unwritten rules, customs, conventions, norms, etc. that govern resource use practices and relevant social relations among the resource users. A co-management institution implies the integration of formal and informal rules and/or government and community actors. This also applies to resource management committees that are made up of government and community actors but governed by formal governmental rules, committees made of community actors but governed by formal governmental rules, and committees made of governmental and community actors and governed through the integration of formal and informal rules.

Actors in this thesis are defined as individuals and organisations of individuals (e.g. farmers' groups) implementing various natural resource use practices and/or involved in management practices. Actors therefore include governmental staff at various levels – i.e. district/local and farmers, livestock keepers and fishers, as well as their organisations – interacting and intervening at various stages in the resource use process. Actors also refer to non-governmental organisations that undertake the role of building the capacity of governmental and community actors in the management of natural resources. I view them as actors that link the community and the government and build the capacity of co-management institutions in the enforcement of sustainable natural resource management rules and practices.

Institutions play an important role in governing the management of natural resources. In the next section, some institutional dimensions will be used to elaborate on the role of institutions in natural resource management.

2.5 Institutional dimensions and their role in natural resource management

With regard to natural resource management in Tanzania, some institutional dimensions are important to consider if one is to better understand the institutions and their roles in the co-management of natural resources. These dimensions include empowerment, scale, participation, heterogeneity, property rights, and leadership (Figure 2.2). In this section, I review the literature on the impact of these institutional dimensions on natural resource management.

2.5.1 Empowerment

Empowerment can have various definitions in the context of natural resource management. Nielsen *et al.* (2004) define co-management as a way of providing people, as natural resource users, with a chance to influence their own future, cope with external impacts, compete in the use and management of natural resources, and address other issues related to natural resource. It gives an individual the ability to influence and change events, courses of action, and outcomes in his or her life. Empowerment occurs through coping and adapting to a given situation or environment based on flexible responses to various influences (Raik, 2002). In other words, empowerment enables the involvement of some institutions that may enable actors to positively influence sustainable natural resource management.

Empowerment and co-management are interdependent and reinforce each other. Raik (2002) terms co-management a cyclical situation in which the partners develop new skills, institutions and behaviours throughout the process, which results in the empowerment of the participants as individuals and as a community. In turn, when the acquired knowledge and skills are reinforced through practices, co-management is further enhanced. Empowerment has several benefits for

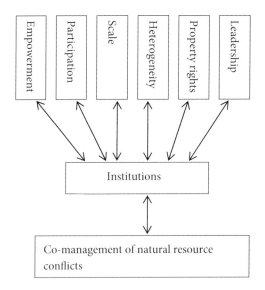

Figure 2.2. Dimensions relevant for understanding the institutions governing the co-management of natural resources in Tanzania.

co-management. First, through empowerment, some institutions may evolve, which may reduce the marginalisation of resource users. For example, Pomeroy *et al.* (2001) argue that empowerment builds the capacity of individuals and groups of resources users economically and politically. In an economic sense, empowerment transfers access and control of natural resources from a few individuals to the poor majority. Through this process, they assert, the community political capacity will be enhanced, which is important for balancing power relations for collaborative natural resource management. Second, the empowerment of individuals and groups is important for co-management. Through this process, access to information is enhanced, desire for change is promoted and enhanced, capacity for controlling and managing natural resources is improved and environmental consciousness may be raised.

Both the government and the local community benefit from empowerment. Empowerment is not a zero-sum game because the actors that participate in the process gain in one way or another. In this process, there is two-way shared learning among actors regarding how co-management can improve problems related to natural resources management (Jentoft, 2004). Nielsen *et al.* (2004) posit that both the government and the local communities need changes in both mindset and skills in order to cooperate and that empowerment can help in achieving the change. Whereas the government may need to outfit its staff with new skills for collaborative management, the community simultaneously may need to develop some capacity to effectively partake in co-management. Empowerment makes individuals and groups/community influence each other. Usually, empowered individuals may influence others in the community through interpersonal interactions. In this process, the community will be undergoing empowerment. On the other hand, community empowerment enhances the empowerment of individuals over the course of the implementation of collaborative practices (Jentoft, 2004). This implies that empowerment can equip the community and individuals in the community with skills and knowledge, which enables them to collaborate with the government in natural resource management. However, this does not mean that people and government institutional actors do not have knowledge at the outset of co-management arrangements.

Both the local people and the government have skills and knowledge related to various issues of natural resource management, and the potential exists for them to enhance these invaluable attributes and develop new ones. The enhancement of skills and knowledge can be operationalised through training to increase, among other things, the knowledge and information levels of those participating in co-management. Training can be offered on leadership, situational analysis and problem-solving, consensus-building, conflict management, etc. External agents, including government agencies and non-governmental organisations (NGOs), may provide this training. This training may help the community to become aware of the resource use problem, make decisions, and assist in preparing the management plans (Pomeroy *et al.*, 2001). Because of the limited capacity of the government in terms of financial resources and limited human resources, and because of the declining power and role of the government in natural resource management, NGOs have become an important partner of the government in empowering actors in co-management initiatives. They play the role of intermediary.

Non-governmental organisations (NGOs) are an important actor in enabling empowerment of those who participate in the co-management of natural resources (Nielsen *et al.*, 2004). They organise, finance, link and empower the community for resource management. Additionally, NGOs mobilise local resource users to form groups (community organisation) through which they conduct capacity-building training (Thompson, Sultana, & Islam, 2003). They also provide credit and support and link local resource users to other microfinance institutions that offer low-interest credit (Thompson *et al.*, 2003; Thompson *et al.*, 1999). They create in local users an awareness of the impacts of unsustainable natural resource management practices, and they assist them in forming institutions and techniques for sustainable natural resource management (Lim, 2006; Thompson *et al.*, 1999).

NGOs can be activists for the interests and rights of vulnerable groups in the community. They can stand up for the rights of the marginalised and subordinated (e.g. women and the poor) and share the benefits resulting from the management of natural resources (Thompson *et al.*, 1999). NGOs also play an advocacy role for grassroots communities, facilitating their acquisition of access rights, and enter into contracts with the government under specified conditions. They bridge a gap between the government and the local community, and they play an advocacy role for the community (McConney *et al.*, 2004). Such advocacy roles include, for example, facilitating collateral-free credit for a poor section of a community (e.g. the landless and those with no assets) and enabling them to initiate environmental friendly projects. The goal is to institute an equitable co-management arrangement between the government and the local community for the purpose of natural resource management (Ahmed, Capistrano, & Hossain, 1997). An NGO can also advocate for the rights and interests of specific local users when, for example, the government wants to divert particular natural resources to powerful users (e.g. external users). In such a case, the NGOs may pay for a lease on behalf of the local resource users, a pre-requisite set by the government for granting use rights (Thompson *et al.*, 1999). They may also train the community to understand rules, laws, procedures, etc., and the use of these instruments to defend their interests (Umar &

Kankiya, 2004). This training may give the community the ability to partake in negotiations with the government and other actors in a co-management process.

NGOs may facilitate the formation of co-management institutions in areas where such arrangements do not exist (Nielsen *et al.*, 2004). This may be useful in reducing unsustainable natural resource exploitation practices. NGOs may do this by establishing links with local elites and leaders who are influential in mobilising sustainable natural resource management (Thompson *et al.*, 1999). The NGOs thus may empower the local community to be able to manage natural resources in a sustainable way using locally available formal and informal institutions.

NGOs can create a partnership with the government that enhances the participation of the community in co-management arrangements (Nielsen *et al.*, 2004). This partnership has resulted in decreased transaction costs associated with the management endeavour (Lélé, 2004). Additionally, where the government is limited financially and in terms of human resources and may thus have trouble reaching many local resource users, a partnership with an NGO helps make those connections (Ahmed *et al.*, 1997, Lim, 2006). The government may establish a partnership with an NGO for various reasons. The action can be geared towards enhancing the involvement of the local users in the conservation and management of natural resources. The intention can be to use the NGO's experience with and knowledge of human development, training, and diversifying sources of livelihood to relieve a specific resource from over-exploitation. It can also be used for capacity-building for local users who will thus be able to partake in decision-making regarding natural resource management with the government (Ahmed *et al.*, 1997). However, when there is distrust, a partnership between a government and an NGO can be strained. The government may see the NGO as a threat to its socio-political position, whereas the NGO may regard the government as being inefficient and corrupt (Lim, 2006).

NGOs in some areas implement functions and roles that are supposed to be covered by the government, especially when the latter is short on funds (Pomeroy, 1995; Figueroa, 2002). However, because the NGOs are similarly constrained by lack of funds, in most cases they have failed to undertake their co-management responsibilities adequately. Although sometimes the government may administratively grant the NGOs the mandate to collect user fees to meet operational costs, the legal implementation of these decisions has been low. This becomes a limitation on the effective operationalisation of NGO responsibilities (Figueroa, 2002).

Although co-management can help ensure success through a partnership between the government and an NGO that empowers actors in co-management arrangements, it can sometimes also lead to negative consequences. Pomeroy *et al.* (2001) see a likelihood of power imbalances in the community because of empowerment programmes. They caution that if empowerment is not carefully implemented, it may create and enhance inequalities in the community, creating a redistribution of power elites. In such situations, instead of reducing social stratification and allowing people to collaborate, empowerment can result in the marginalisation of some resource users in the community. This may further result in conflicts within a community and between the community and the government.

2.5.2 Scale

Scale is defined as a level of geographical resolution on which decisions are thought of, worked on or studied (Agnew, 1997). Scale is socially constructed, and therefore, besides implying size and levels, it also reflects social relations. In this case, it reflects a context of social empowerment, disempowerment and physical environment in which social processes takes place (Howitt, 1998; Marston, 2000). Cash *et al.* (2006) define scale as spatial, temporal, jurisdictional, and institutional dimensions that can be used to measure or study a given phenomenon or event. These authors argue that two or more scales can interact or link to influence certain phenomena; thus, the term 'cross-scale linkages' is used to connote such a situation.

Co-management is an example of a system that forms cross-scale linkages. Linkages that result from co-management are crucial for providing information used by actors (who otherwise would lack that information) to deal with or to organise themselves to deal with certain challenges or problems (Lebel, Garden, & Imamura, 2006). Cross-scale links become especially important where governing institutions at various organisational levels are weak (Barrett, Brandon, Gibson, & Gjertsen, 2001). Even when local institutions are relatively strong, they cannot enforce sustainable natural resource management on their own because they are embedded within larger government institutions and legal and policy environments. These influence the local institutions and the resources under question (Berkes & Seixas, 2004).

Cross-scale interactions may involve horizontal levels or vertical and horizontal levels. Horizontal cross-scale interactions occur among local users, such as between upstream and downstream users of water resources or between departmental levels in a district. Vertical cross-scale interactions occur between the government – e.g. at the district level – and the local community (Senyk, 2005). This occurs, for example, when the district authorities are involved in resolving conflicts among resource users at the local level. Cross-scale interactions may also occur on a temporal scale: for example, during certain seasons of the year (Wilson, Ahmed, Siar, & Kanagaratnam, 2006).

Berkes (2002) asserts that strengthening the local institutions is not by itself a guarantee of effective co-management arrangements. For effective co-management to happen, reconciliation between government (top-down) and the community (bottom-up) institutional arrangements is needed. For this reconciliation to occur, Berkes claims, cross-scale management systems have to be devised. These management systems enable horizontal institutional linkages among user groups and geographical areas (Jentoft, 1999), and vertical institutional linkages across local users and political scales outside the local area (Downie & Fenge, 2003). It is quite often impossible to find a resource management system without cross-scale institutional linkages and drivers at different scales (Berkes, 2002). Berkes (2003) and Berkes and Seixas (2004) underscore the importance of cross-scale institutional linkages as a way of sharing information on the status of natural resources and constraints to natural resource management. For them, based on cross-scale linkages between the government and local resource users, strategies can be devised to ban unsustainable natural resource use practices such as the use of unsustainable fishing gear.

The cross-scale nature of natural resource management problems implies that cross-scale institutional solutions are necessary to solve them. In this sense, community institutions and actors alone cannot adequately resolve these problems. For the same reason, centralised (state-based) institutional systems alone cannot adequately provide solutions to natural resource degradation issues (Carson & Berkes, 2003). Berkes (2002) asserts that even for centralised natural resource systems, the government needs the local people – for example, it might need the knowledge and skills of the local users. Thus, cross-scale institutional relationships are imperative for addressing complex situations by pooling together government and the community institutions and resources (e.g. knowledge, skills, and information-sharing) (Olsson, Folke, & Berkes, 2004). In this way, the scholars claim, problems beyond the capacity of one actor can be collaboratively resolved.

Other scientists view cross-scale institutional relationships between the government and the local community from a transaction-cost point of view (Carlsson & Berkes, 2003; Solecki, 2001; Wilson *et al.*, 2006). According to these researchers, cross-scale interactions can lower transaction costs among local resource users. They also claim that co-management arrangements create webs of relations and links for providing information, addressing legal issues of resource use, and monitoring the use and practice of resource users. These links and webs can mediate property rights issues and hence may result in lowering transaction costs in the use of the resources. Likewise, the authors claim, a cross-scale institutional relationship can be useful in resolving conflicts among the users of the resource. This occurs when, for example, a government involved in a co-management arrangement controls the right to resource appropriation on the part of local users. The likelihood of conflict among resource users thus is minimised, and in turn, transaction costs in terms of the time that might have been used in resolving these conflicts are lowered.

Olson and Folke (2001) view cross-scale institutional interactions as platforms for assessing the effectiveness and feasibility of formal mechanisms and instruments devised by governmental actors but implemented by community actors. In such interactions, the local people, entrusted with the enforcement of resources management rules, may provide feedback regarding issues and challenges encountered in the enforcement process. This collaboration of the government and the community is an indication that the participation of various institutional actors is important for addressing resource management problems because individual actors cannot adequately resolve resource use and management challenges.

2.5.3 Participation

Cross-scale interactions imply the participation of various actors in negotiating or addressing issues related to natural resource use and management. In this section, I analyse the interplay between co-management and participation.

The participation of formal/informal actors and government/community actors in the management of natural resources is essential to the sustainable management of natural resources. The participation of these actors implies the existence of interplay between the formal and informal institutions that govern the practices and decisions of actors in natural resource management in a social-ecological system. The government and the community, besides enforcing sustainable management practices, may collaborate in formulating rules and regulations that may govern and improve their management practices and decisions. People around natural resource systems should participate in the formulation of resource use and management rules because changing rules governing resource management may influence their resource use patterns (Pomeroy, 2001).

Community participation in rule formulation with the government and other stakeholders should not end with choosing the rules but should also extend to implementation and enforcement. Community participation in the enforcement of the rules gives them a chance to understand how the rules operate in practice and what issues may determine their feasibility. In addition, it increases the community's awareness of the constraints of the existing rules. Thus, they will be in a position of proposing and participating in the revision of the rules to address these constraints (Isaac *et al.*, 1998). The belief behind this assertion is that any policy, even if made by the government, can only become successful when the user community is involved in its implementation.

The interaction of formal and informal institutions in co-management arrangements is imperative for sustainable natural resource management (Lingard, Raharison, Rabakonandrianina, Rakotoarisoa, & Elmqvist, 2003; Pomeroy *et al.*, 2001; Yasmi, Colfer, Yulian, Indriatmoko, & Heri, 2007; Wanitpradit, 2008; Naugran, 2002). Integrating informal (customary) institutional frameworks into formal institutional frameworks is especially important when these frameworks have to do with the goals, purposes and outcomes of a management initiative. In such collaboration, shared and adaptive learning will occur, which can lead to the building of social capital for sustainable resource management (Baland & Platteau, 1996; McCay, 2002; Plummer & FitzGibbon, 2007; Schusler, Decker, & Pfeffer, 2003). The collaboration of formal and informal institutions can rescue a resource that both the government and the community would independently fail to protect. Naguran (2002), for example, highlights the case of the Ndumo Wetland Reserve in South Africa, where government institutions and local people formerly failed to address development and conservation objectives but have finally developed a partnership involving formal and informal institutions whereby formal and informal actors participate in negotiation and rights reformation to address the situation. Working on a similar case, Glaser & da Silva Oliveira (2004) report how a government-controlled mangrove system resulted in open access when the Brazilian government could not enforce sustainable management. However, when co-management was instituted, sustainable management was restored again. Conservation was possible when the formal governmental institutional mechanisms recognised and encouraged the participation of the community and integrated informal community rules, thereby creating collaborative institutional arrangements.

These cases indicate that it is difficult to attain conservation and development objectives where formal and informal institutions operate in isolation. The integration of the government and community institutions seems an appropriate option because the problems seem to demand the capacities of many independent institutional actors if they are to be properly addressed.

Collaboration between formal and informal institutions may be important for the enforcement of sustainable natural resource management initiatives. In order to be enforceable, the traditional natural resource enforcement rules may need to be supported by government institutions. In the same token, government institutions may be useful in enforcing rules when actors and users from outside the community are used in the management or use of natural resources and where the inside formal or community rules are not positioned so as to restrict the influence of external users. Therefore, participation may occur in terms of integrating formal and informal institutions to regulate the unsustainable use of natural resources (Pomeroy *et al.*, 2001).

Informal and formal institutions share jurisdiction in natural resource management. Some scholars (e.g. Wanitpradit, 2008) claim that the introduction of a co-management arrangement creates cooperative natural resource management between the state and informal management regimes. In the integrated arrangement, some responsibilities may be shared, while others may remain under informal or formal jurisdiction. Leaders for co-management institutions may be selected based on their awareness and knowledge of governing institutions, including formal rules like policies and laws or custom-based institutions like rituals, customary rules, conflict management institutions under clan and kinship systems. Informal institutions can be a source of useful traditional skills, knowledge and experience in a co-management system (Carter, 2008).

Spielmann and Unger (2000) assert that the wisdom, patience, and experience of elders can be integrated with formal knowledge to improve decision-making and enhance the success of co-management. In some cases, however, co-management can erode traditional institutions that existed before its instigation (Langard *et al.*, 2003; Njaya, Donda, & Hara., 1999). The power and influence of traditional leaders may be eroded. Some scholars (e.g. Thomson & Gray 2007) assert that the superimposition of a resource management policy by the government on the community without consideration of existing management systems and rights in the local area ultimately undermines useful local institutional arrangements and marginalises poor people. The erosion of the power of informal institutions, nonetheless, may have positive or negative implications. If the informal institutions were unjust, unfair, and discriminative, such that some individuals accumulated power and resources in as selfish way, their erosion might contribute to creating equitable community. However, the erosion of good institutions might render co-management ineffective (Njaya *et al.*, 1999).

This indicates how complex the interplay of informal/formal and governmental/community co-management arrangements, institutions and actors in a social-ecological system can be. Whereas one institutional actor can assist in reducing or eliminating undesirable elements, consequences and weakness as related to another institutional actor, the desirable and useful attributes of one institutional actor can also be undermined and eroded by the other.

Governmental and community actors may actively participate in the co-management of natural resources when trust builds among them (Taiepa *et al.*, 1997). Trust involves the mutual recognition of the learning patterns of cultures participating in co-management (Kendrick, 2003). Trust is a mutual process that can enable the involved parties to develop solutions to problems that are difficult for one institutional actor to solve (Pomeroy *et al.*, 2001). Co-management may enhance trust-building among the participating actors and stakeholders. The government (using its research agencies) and the users of a particular natural resource may perform joint action to investigate the endangerment of a species or the impact of the use of certain harvesting gear on the sustainability of the resource (Kaplain & McCay 2004). This may enhance trust between the government and the users of resources through the sharing of mutually important problem investigation and solution implementation processes.

However, the development of trust in co-management may take time. In some cases, the resource users might not trust the government, thus requiring time to develop trust through the creation of platforms for discussion and dialogue among the partners that allow open, frequent and ongoing communication and dialogue (Pomeroy *et al.*, 2001; Taipea *et al.*, 1997). In these dialogues, important aspects such as power-sharing, commitments, and benefit- and cost-sharing should be discussed, while accountability and enforcement mechanisms should be clearly defined.

There should also be discussion of mechanisms for enforcing agreements and of sanctions to be imposed for non-compliance (Pomeroy *et al.*, 2001).

In principle, the literature in this section can be summarised as follows. Scholars agree that the integration of formal and informal institutions and the participation of governmental and community actors is important for co-management. Such integration and participation can assist in rescuing natural resources, which are deteriorating where individual institutions and actors (e.g. formal or informal alone and/or governmental or community alone) have failed to address the problem. Whereas participation can assist in eliminating the impacts of and erode some negative institutions, it can also result in undermining and eroding useful institutions. Through participation, some good institutions may be eroded, especially if the imposition of new institutions in an area does not determine the power of the existing institutions.

Up to this juncture, participation has been analysed within the framework of a government-local community dichotomy. Such a dualist view is likely to mask the heterogeneous nature of the community and the government. In practice, participation entails diverse and heterogeneous community groups and diverse government units governed by diverse and heterogeneous formal and informal institutions. This diversity has implications for the co-management of natural resources. In the coming section, the influence of heterogeneity in the collaborative management of natural resources involving governmental and community institutions and actors is analysed.

2.5.4 Heterogeneity

Although heterogeneity is a term with many meanings based on the context, in the present thesis it implies the existence of multiple and diverse user groups, either permanently or on a seasonal basis in a given area, with the aim of exploiting natural resources. The main dimensions referred to here are ethnicity and interactions between resident and non-resident resource users, as these seem most relevant for the Lake Jipe case study.

Ethnicity in this thesis implies the presence of resource user groups that, though they may be implementing the same practices (e.g. fishing, livestock-rearing, etc.), have different cultural origins. In light of the existence of multiple ethnic groups, natural resources governing institutions at the community level may disintegrate along ethnic institutional lines. The relationships between the local user groups and between those groups and the government may be determined by the cultural institutional relationships of a given area. Thus, cultural diversity may have implications for natural resources management.

Social communities are diverse entities embodying different expectations, ethnicities, thoughts, perceptions, objectives, interests, levels of wealth, etc. (Natcher, Davis, & Hickey, 2002). Ethnicity is among the important aspects in the co-management of the natural resources. It is an attribute that indicates that a resource user's community is variable (Sick, 2002). Along with gender, class, and caste, ethnicity is a basic dimension of conflicts in natural resource management (ICLARM, 2001). Understanding the rights and interests of various ethnic groups and multiple governing ethnic institutions in particular resource settings may be imperative for planning collaborative natural resource management (Armitage, 2005).

The ability, capacity and willingness of different ethnic groups may affect natural resource management arrangements, either positively or negatively. The existence of multiple diverse ethnic groups characterised by different values, expectations, aspirations, and interests as relate to natural resources may result in conflicts among them (Armitage, Marschke, & Plummer, 2008). On the other hand, cultural diversity, differences in perception and understanding, and reciprocal relationships imply the existence of diverse solutions to problems of constraints. In other words, they mean the existence of diverse social networks and human capital. Strategists and practitioners of natural resource management may need to clearly understand these cultural diversities if the outcome of the co-management of natural resources is to be improved. Such understanding may enable the mobilisation of social and human capital through the integration of formal and informal actor networks and institutions and consequently may improve natural resource management outcomes (Jentoft, McCay, & Wilson, 1998; Stratford & Davidson 2002).

Ethnicity can contribute to the success or failure of a co-management system. Some scholars emphasise that in a situation in which homogenous ethnicity exists there is a greater chance of attaining successful co-management than for heterogeneous ethnicity (Jodha, 1996; Lim, Yoshiaki, & Yukio, 1995; Pinkerton, 1987). Kideghesho and Mtoni (2008) reveal that heterogeneous ethnicity results in the delayed development of co-management institutional arrangements. In their research, these authors found that the presence of many ethnic groups within a particular social-ecological environment results in dilution effects for the culture, leading to a lack of social cohesion; they also discovered that one strong traditional institution is required for building an operational local organisational setting as a preparatory phase for co-management.

For other scholars, nonetheless, the success or failure of co-management arrangements in light of ethnicity depends on whether the existing cultural institutional diversities are subverted or engaged. Natcher *et al.* (2005) claim that when differential cultural groups subvert their cultural diversities in resource management, the result is that co-management arrangements are undermined. Diversity might be used to fuel struggles over resources or as a stage for future conflicts (Sneddon, Harris, Dimitrov, & Özesmi, 2002). Instead of collaborating, resource users might compete for socio-economic purposes and interests, thus sustaining and even escalating conflicts among them. On the other hand, if the cultural institutional diversity (e.g. cultural knowledge and experiences) is engaged, there exists a greater chance of developing a successful co-management arrangement.

As we have seen above, different ethnic groups may sometimes coexist harmoniously within the same social-ecological environment and implement sound natural resource use and management practices. In such cases, the influence of ethnicity on co-management may be a lesser priority. The primary concern for co-management may be any existing tension between the resident users (insiders) and non-resident users (outsiders) of the particular resource.

The relationships between resident and non-resident users may determine the nature of the relationships between the government and the local resident users and formal and informal institutions in the management of natural resources. The resident users may be motivated to form a co-management arrangement with the government to defend their interest in a resource, which they may think could be compromised by non-resident users. The community, through collaboration with the government, may gain some authority that it may use to exclude outsiders who are not welcome in the local area (Nielsen *et al.*, 2004). The use of more advanced fishing gear by non-residents might be a rationale for the community's seeking authority from the government because the community regards these gears as a threat to realising the sustained availability of the resource to provide income and food security to residents (*ibid.*).

For some cases, nonetheless, the influence of non-resident resource users disintegrates collaborative institutional arrangements, not only between the government and the local users but also among the local user groups. This is when the influence of non-residents results in a conflict of interest among partners in co-management arrangements. While some may want to adopt advanced technology to enhance the exploitation of resources for commercial purposes, others may wish to maintain traditional harvesting technology, adequate for exploiting the resource for subsistence purposes (Sekhar, 2004). This may lead to conflicts and resource degradation because the stability of the governing institutions may be impaired. Some scholars (e.g. Castello, 2001) claim that the solution in such a situation is to provide exclusive access rights to the local people; through this institutional adjustment, they argue, non-resident users (outsiders) will be excluded. However, sometimes, Sekhar argues, the government may be in favour of one side – e.g. it may promote investment in an advanced technology to enhance revenues from taxes. Under such circumstances, with the government now partial to one side, it may hardly be in a position to resolve conflicts between users, and this may undermine sound institutional arrangements developed for collaborative and sustainable resource management.

In the above section, we have reviewed how heterogeneity influences co-management. Two dimensions related to heterogeneity – i.e. ethnicity and interactions between resident and non-resident resource users – have been highlighted. Ethnicity can have positive or negative influences on co-management, as we have seen, depending on whether diversity is engaged or subverted. The literature has also indicated that interactions between resident and non-resident resource users are in most cases conflicting and may have negative implications for natural resources. One strategy for dealing with this challenge, as suggested in the above account, is to provide users with exclusive access rights to exclude outsiders. In the coming section, I turn to investigating how property rights may influence co-management.

2.5.5 Property rights

Jentoft (2005) describes property rights as relations between people, the owner of a property or good and a non-owner, in terms of their relative positions. In this case, the owners have the legal right to deny non-owners the enjoyment of benefits from the property. Jentoft asserts that the relationship is not between the property and the owners but rather between the owners and non-owners. This implies that there are institutions that govern and determine the interactions between the owners/non-owners and the resources. It also suggests that both the owners and the non-owners are aware of these institutions. These institutions may define actors entitled to use the resources and temporal and spatial boundaries of use, as well as conditions for use.

Several property rights may exist, including private, state, open-access, and communal rights, or even combinations of these. Private property rights are adopted where there is limited land, the frequency of its use is high, and the population is high; thus, the type of property rights becomes strict and more explicit. On the other hand, where land is abundant, population density is low, and there is extensive utilisation of land, the property rights adopted are often communal and less strict in the sense that shifting is possible. In some areas, a combination of private and communal property rights may exist, and in other areas, what is considered communal may be a complex mix of individual ownership (e.g. maize fields) and group ownership (e.g. grazing land) (Wiber, 1993).

The type of property rights under which a co-management system operates determines how strong the co-management arrangement can be. A co-management arrangement that operates under communal property rights has the power to control access to the resource, impose sanctions in response to non-compliance, and ultimately enforce exclusion if the expressed areas of dissatisfaction are not addressed. With this property right system, it is possible not only to condemn unsustainable use of a resource contrary to an agreement but also to exclude actors who do not abide by the governing institutions. On the other hand, for state, private and open property rights systems, co-management arrangements operating under them do not have a power to sanction by exclusion. Individuals or institutional actors who do not abide by the agreed-upon regulations have an exit option. If they do not want to negotiate for collective action, they may decide to make their own decision. Co-management arrangements under these four property rights circumstances can only have the moral power to condemn non-compliance; they cannot impose sanctions through exclusion (Jentoft, 2005). Despite these power differences, nonetheless, Jentoft claims, co-management can succeed under any of the mentioned property rights arrangements. This is an indication of the existence of other attributes that may influence the relationship between co-management and property rights in the management of natural resources. One such attribute might be the existence of a legal mechanism.

Pomeroy *et al.* (2001) assert that the existence of property rights alone is not sufficient to enforce the sustainable use of natural resources. Whereas property rights should clearly provide for mechanisms (administrative, economic, and collective) and structures for the allocation of property rights to optimise their use while conserving resources, there should also be a legal mechanism for enforcing rights (Pomeroy, 1995; Pomeroy & Berkes, 1997). These scholars posit that the government should be legally able to support the enforcement of property rights to ensure, for example, that local people have the power to control the unsustainable use of natural resources by outside pressures. This is not to say, however, that when the two attributes (property rights and legal support) are appropriately in place, then the success of co-management arrangement is totally guaranteed. The success of co-management is determined not just by the existence of property rights and legal support but also by the interactions among various formal/informal and government/community actors in enforcing these rights, as well as by how the existing institutions and arrangements in a given social ecological environment govern the actors in enforcing these rights in their practices and interactions.

2.5.6 Leadership

Other important dimensions influence co-management as well. One such dimension is leadership. In this section, I review the influence of leadership on co-management. Leadership is an important condition for the success of co-management. Whereas some scholars (e.g. Pomeroy *et al.*, 2001) have analysed the roles of leaders within the resource user community on co-management, others (e.g. Singleton, 2000) have analysed the roles of state/government leaders in co-management. These scholars agree that leadership is an important element of the success of co-management arrangements. Leaders portray examples for others to follow; they show the way and mobilise energy and power for the co-management process. While the community might already have leaders, they might not be the appropriate ones for co-management arrangements (Pomeroy *et al.*, 2001).

With their focus on leadership from the community point of view, Pomeroy *et al.* (2001) assert that leaders should be chosen from among the users by the users themselves in order to gain legitimacy and respect from their counterparts. However, they also highlight that there must be institutional provisions that clearly define the time period during which one can remain a leader so that no one maintains the position in the long term. Other community institutional actors should also be given an opportunity to develop leadership skills. While, on the one hand, periodic change of leadership might be useful for avoiding over-dependence on a few institutional actors, which might become a problem when they are not available, it also reduces the chance for corruption on the other hand. Furthermore, these scholars assert, the over-dependence of one individual might become a constraint when that person dies, leaves his position or moves to another because the user's community might at that time not have developed other people to take his place. However, regarding the appropriate duration of a leadership position in the community, there are questions to consider. What length of term reduces the chances of corruption? How can this figure be determined and operationalised, and who determines it? Perhaps short-term leaders might be more corrupt than long-term leaders depending on the context.

Other scholars, as introduced above, have determined the influence of leadership on co-management from the perspective of the government. Singleton (2000) highlights the role of government leaders in co-management arrangements. She asserts that government leaders are essential facilitators of cooperative decision-making between the government and the community. They can do this by creating opportunities for face-to-face interactions through which shared interests can be explored outside the formal institutional context of decision-making. In such informal interactions, Singleton argues, skilful leaders can have the opportunity to create occasions for cooperative covenants between the community and government.

The above account in this section implies that collaboration between formal/informal, and government/community institutions and actors is important in developing leaders for co-management of natural resources. This collaboration may be governed, enabled and enforced by governmental and community institutional mechanisms. The collaboration of formal and informal institutions may determine success or failure in developing these collaborative leadership skills and actors in co-management arrangements in the social-ecological system.

2.6 Conclusion

So far, I have defined the concepts used for analysing the co-management of natural resources in the Jipe social-ecological system along with factors that characterize the institutions – that is, the institutional dimensions. However, the Jipe social-ecological system is a complex system wherein there exist at least three arrangements and where, therefore, multiple government and community and formal and informal institutions govern the practices of multiple and heterogeneous actors who play diverse roles within it. The three arrangements are fishery, agriculture and livestock sectoral management systems (Figure 2.3).

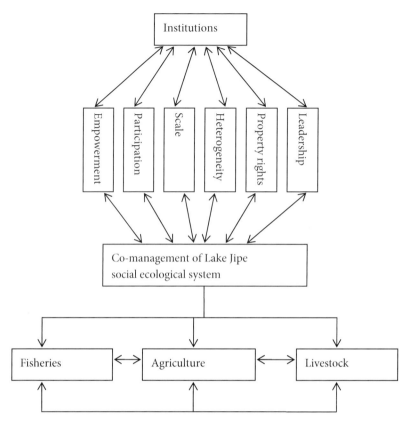

Figure 2.3. A complex lake Jipe social-ecological system with three interacting co-management arrangements: fisheries, agriculture, and livestock.

The analysis focusing on these three different natural resources co-management arrangements within the same social-ecological system goes beyond the conventional analysis that focuses on co-management in individual natural resource management systems. The analysis in this thesis will provide insights into and challenges to co-management theory and enhance our understanding of how diverse but interdependent internal sectoral resource management politics within the same social-ecological system can affect the co-management and sustainable management of natural resources in the social-ecological system.

In the next three sectoral empirical chapters, I analyse the co-management arrangements with multiple governing formal and informal institutions and multiple government and community actors in the management of natural resources in the Jipe social-ecological environment. Examining one of these co-management arrangements, we will see how a non-government actor (NGO) links and builds the capacity of the government and the community actors to enforce sustainable natural resources management practices. Prior to these chapters, however, chapter three will give the history of the governmental-level institutions in charge of natural resource management in Tanzania.

Intermezzo – Lake Jipe: introduction

This section introduces lake Jipe and describes the methodology and methods used to collect data for this thesis. The section first gives a short description of the study area, including the location and climatic conditions of Lake Jipe, and the socio-economic, ecological, and cultural characteristics of the region. Subsequently, data collection methodology is highlighted and the methods, instruments and tools used under this methodology are presented. This entails the use of a combination of methods for complementing and confirming data collected using other methods. The section ends by introducing methods used for data analysis.

1 Study area

The study was conducted around Lake Jipe, an area encompassing upstream and downstream wards of Jipe, Kwakoa, and Mwaniko in Mwanga district. Lake Jipe is a shallow lake located on the Tanzania-Kenya border between 3°31' - 3°40' S and 37°45'E. The lake covers an area of 30 km², and is 12 km long, 3 m deep (Ndetei, 2006), and 2 km wide (Dadzie & Haller, 1988), and is on the leeward side of north Pare Mountains (Ndetei, 2006; Twongó & Sikoyo, 2001). Climatic conditions of lake Jipe are sub-tropical semi-arid conditions wherein rainfall ranges from 500 to 600 mm, and temperature ranges from 19.9 °C to 29.5 °C. While rainfalls are of biannual patterns with long rains from March to May and short rains from October to December, the hottest months are January to March April and the coldest month is September (ESF, 2005).

Lake Jipe is an important field on which various natural and anthropogenic activities and interactions exist including ecological, socio-economic, cultural and political interactions as the underneath sub-sections unveil.

1.1 Ecological importance

Lake Jipe is connected to Lumi river and other streams originating from Mount Kilimanjaro, and to Muvulani river and small streams including Kirurumo from Pare Mountains. These river systems are sources of water to this lake (IUCN, 2000). Lake Jipe outflows into Ruvu river which is a tributary of Pangani river and therefore lake Jipe is the storage basin of the Pangani basin (Ndetei, 2006). Lake Jipe is rich in biodiversity by having water-birds: Lesser Jacana, Purple Gallimule, Squacco Heron, Black Heron, African Darter, African Skimmer and fish species namely endemic Tilapia, *Oreochromis jipe* and sardine, *Rastrineobola argentea*. In addition, it is a habitat of crocodile and hippos (Twongó & Sikoyo, 2001; MNRT, 2004). However, the lake, which originally covered an area of 100 km², has been reduced to 30 km² today due to unsustainable resource use practices including catchment degradation, leading to erosion and siltation as well as the rapid growth and expansion of waterweeds of *Typha domingensis* (Ndetei, 2006).

1.2 Socio-economic value

Jipe wetland provides livelihood support to many people on both sides of the Kenya-Tanzania border (IUCN, 2000). On the Tanzanian side, around 17,800 people inhabit the lake area (Census

2002). The lake is used for fishing, and provision of water for domestic consumption. In recent years, nonetheless, fish production has been declining due to among others invasion of *Typha* waterweeds, therefore adversely affecting livelihoods of the people who depend on fishing (Ndetei, 2006). Besides fishing, the areas surrounding this lake are used for agricultural crop production, and for livestock keeping. Various crops including maize, beans, cowpeas, lima beans, green gram, sunflower, and tomatoes are grown. The livestock kept in the area include poultry, goats, sheep and cows. Although farming, agriculture and fishing are the main livelihood activities, secondary activities are also carried out including food vendoring, shop transactions, selling of industrial and local beers, selling of firewood demanded for roasting the fish and bicycle repair. All these secondary activities are largely linked to fishing. It creates temporal interactions of residents and non-residents, the latter being more active during fishing seasons. As such, while an excessive fishing pressure has a potential for degrading fisheries resources especially due to unsustainable fishing practices, on the other hand interactions between residents and non-residents create employment opportunities to other inhabitants who do not directly engage in fishing. These inhabitants provide economic goods and services needed by those directly engaged in fishing.

Apart from initiatives by individual villagers and by communities to support local livelihoods, some support and initiatives derive from the state. While this study was being conducted two state driven programmes were being facilitated by the district government in the area. Whereas one programme was not sectorally specific, the other programme was sectorally specific. The non-sectoral specific programme was implemented under the Tanzania Social Action Fund (TASAF), which provides funding support to groups of villagers or village communities as a whole to undertake participatory projects of their choice under the facilitation of the government. The sectorally specific programme was implemented through a funding from the Ministry of Agriculture under the Participatory Agricultural Development and Empowerment Project (PADEP), which targeted groups of farmers who were advised to design and propose participatory agro-projects. In general, villagers in lake Jipe area desire to capitalise on these potential funding arrangements to improve irrigation potentials of the area, which is semi-arid and experiencing recurrent droughts and unreliable and unpredictable rainfall patterns.

Besides being important socio-economically to the inhabitant of lake Jipe and to those at the Tanzania-Kenya border, lake Jipe is important for the district and regional economy. This lake is a source of water to the hydro-electric power generating plant at Nyumba ya Mungu dam, which is a source of electricity (Mtalo, 2005) to not only Mwanga district but also to the surrounding districts.

1.3 Cultural characteristics

Although culture can be used in broad sense, in this context I refer to culture in terms of customs and traditions including some normative practices in the community. Lake Jipe area experiences a mix of interactions of residents and non-residents at different spatial and temporal points. While the main inhabitants of this area are resident Pare ethnic group, other migratory groups of resources users include the Maasai livestock keepers, and fishers from southern, western, central and other nearby regions of the northern zone of Tanzania, which exploit pastoral and fisheries resources at different times and spaces. As such, Lake Jipe is home to interactions of different cultural groups. Although, these cultures remained separated for quite some time, there in a growing tendency of exchange and intermingling of cultures of different groups. Some immigrants to Lake Jipe have intermarried to Pare community, contributing to a gradual transformation of traditional practices and customs in their now culturally hybrid families. For example, while elders revealed that in the past forty or more years residents of Jipe did not engage in off-shore fishing but instead fished using long sticks while standing at the shore of the lake, today Jipe residents also fish within the water lake using wooden canoes. By the same token, while traditionally women at Jipe were not allowed to enter the lake, this tradition has gradually disappeared. In the present days women paddle canoes and get in the lake to fetch water.

1.4 Thesis approach

In this thesis Lake Jipe is regarded as a representative example of a Tanzanian wetland ecosystem, from which more general conclusions on wetland management in Tanzania can be drawn. I have employed a geographic approach in which both social and ecological systems that potentially influence management of Lake Jipe wetland have been integrated in the investigation. Within one geographic area, the existing arrangements of natural resources management – and their interdependencies – have been investigated, focusing on fisheries, agriculture and livestock. Within each arrangement, the institutions and actors that comprise these arrangements have been identified and assessed.

2 Data collection

This study had to rely mostly on the primary information due to the limited availability of official statistical information, reports on Lake Jipe, and earlier academic studies. The reliance on especially primary data collection was time consuming and required a mix of qualitative and quantitative data analysis techniques.

Prior to the actual data collection process a pilot survey was conducted. After revising the research instruments based on the reality and facts acquired at the pilot survey, data collection was conducted. In the downstream area Jipe, Kambi-ya-Simba, and Butu villages were selected from Jipe Ward, and Kigonigoni village from Kwakoa ward, In the upstream area, Mangio, Vuchama-ngofi, and Masumbeni villages were selected, from Mwaniko Ward (Figure 2.4).

Multiple sources of evidence add validity and reliability to research findings (Yin 1994). As such, a combination of instruments were used for data collection. These included a household questionnaire, interviews with key informants, observation, participant observation, and informal talks with individuals. These various techniques targeted relevant actors from village, ward, district, and national levels. This multiple method, multiple level approach make up the primary data collection methodology. In addition, secondary data were collected through documents at various offices (village, ward, district and national), and through internet search.

3 Pilot survey

The Jipe and Kwakoa wards, on the downstream, and Mwaniko ward, on the upstream, were surveyed prior to the actual data collection. These wards surround Lake Jipe, so its inhabitants

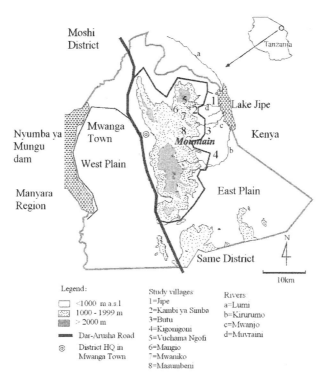

Figure 2.4. A map showing the studied villages of lake Jipe on the upstream and downstream areas.

exploit Lake Jipe and terrestrial natural resources around it. The aim of this pre-survey was for the researcher to familiarize with the reality of the local situation, an imperative step for making decisions on sites/villages for data collection, type of users to include, and other stakeholders from whom to collect data and information.

Natural resources including the lake, farming land, grazing land, and water tributaries on the downstream and upstream areas were surveyed through a cross-sectional transect walk from downstream to upstream. During this cross-sectional pre-survey, a local informant who had a good knowledge on use and management practices of the pre-surveyed areas provided accompanied the researcher. A pre-assessment was conducted to evaluate what natural resource use practices and institutions exist around Jipe. The aim of this exercise was to get an overall picture on status, trends and dynamics on natural resources use and management practices, and coping strategies in the area. In doing so data collection instruments (questionnaires and checklists of questions) were also pre-tested and adjusted accordingly. The pre-assessment provided a general overview important for selecting relevant sources of data, and modifying the questions and instruments (i.e. questionnaires and checklists of questions) to reflect the field reality.

In addition, the pre-assessment allowed for the selection of target villages for data collection. In some villages, for example, a pre-survey revealed that livestock keeping was given more weight than farming (e.g. Kigonigoni village) whereas in other villages (e.g. Butu), more or less balanced integration of different livelihood practices was adopted. Again, for some villages (e.g. Kambi ya Simba), irrigation farming was being conducted, while for others (e.g. Jipe) irrigation was not possible because of lac of available water. This inherent variability in the area is therefore an important selection criterion for the rest of the research.

4 Primary data collection

4.1 Household questionnaire

A household questionnaire was used to collect data at household level. Using this questionnaire questions were asked about natural resources use and management practices in livestock, agricultural production, and fisheries; on reasons for choices or preferences of certain practices; on constraints encountered in the implementation of livelihood practices; on awareness on environmental degradation; on coping strategies for the constraints encountered; on involvement in formal and informal (co-)management institutions; and on conflicts emanating in the use of these resources among different users.

Representative households active in livestock keeping, fisheries and crop production were then purposively selected for further investigation. For each category, random sampling was used for selecting households for interviews. At least 10% of the households were included in the household interviews for each selected village (following Grinnell, 2001). A total of 150 questionnaires were administered. A sample questionnaire that was used for data collection is indicated in Appendix 1.

4.2 Focus group discussion

This method was used to collect information from natural resources management committees in the local area such as fishers committee, the village environmental committee, village water committee, elders committees, informal livestock keepers committee, and a group of fishers. These focus group discussions provided information about the current natural resources situation in comparison with the past situation, problems the committees and institutions encounter in enforcing their management roles, and how they address or resolve them. This information complemented and clarified the data collected through the household questionnaire administration. The groups and committees for focus group discussion were identified during the pre-survey. A total of seven focus group discussions were conducted and an average of six members participated per focus group discussion. Checklists of questions were used to guide the discussion between the moderator and the respondents as indicated in Appendix 2. These questions were used as a guide, and were followed by the follow-up and probing questions through which more detailed information was collected.

4.3 Key informant interviews

The key informants interviewed involved technical and administrative personnel from the village, ward, district, up to the ministerial level. At the village level, the village executive officer, village agricultural extension officer, ward agricultural extension officer, ward executive officer (WEO) were involved in the interviews whereas at the district level, district natural resources officer, district fisheries officer, district agricultural and livestock officer were interviewed. At the ministerial level, staff at the Department's of fisheries, livestock, and agriculture were interviewed. At the local level the key informants were questioned as to their experiences in enforcing natural resources management institutions, and constraints they encounter in the process. At the district level, the key informants were questioned on how they facilitate sustainable natural resources management practices through formulation of by-laws, strategies, etc.; on constraints and challenges they encountered; and on how they resolve them. Likewise, an investigation was made on how the district government collaborates with the central government in the implementation of the relevant institutional arrangements (e.g. policies, acts, strategies, etc.). At the national level, an assessment was done on how the government policies and strategies accommodate and involve the local users in their formulation. A checklist of questions asked to key informants is provided in Appendices 3a, b and c. A list of the interviewed key informants is included as Appendix 4.

4.4 Observation

Observation supplemented and/or confirmed the data collected using other approaches. The researcher was accommodated at the residence of the Ward Executive Officer (WEO). This residence was both used as an office and the house for this official. The ward office is used as a local tribunal in the Tanzanian legal framework. Various issues, socio-economic and emanating along the resource use processes, are reported to the WEO if the lower government and the community instruments have failed to resolve them. Since the researcher was there when these issues were being reported and addressed, he could be informed and learn what was going on. For example, through this approach, it was possible to uncover that some village leaders partake in natural resource degradation practices, or illegitimately protect unsustainable natural resources use practices. These government leaders were reported to the WEO by leaders of informal institutions. Though the same village leaders had been earlier interviewed, it was not evident yet until the researcher observed arguments between the two sides occurring at the WEO's office. The disadvantage of staying at the WEO's office might be that the investigator was perceived to be allied with the WEO, potentially reducing openness of respondents in other data collection activities. However, no indications of such bias appear in the analysis.

4.5 Participant observation

For some farmers, interviews were conducted on their fields because at that time they were engaged with fieldwork. This was also important because along with verbal responses, the researcher could see what was, in practice, transpiring in the field. This activity occurred during July and August at the time of crop harvesting (e.g. maize) and preparation of fields for planting other crops (e.g. beans). Data collection in the field therefore combined different approaches including questionnaire administration, informal talks during work and formal discussions afterwards. The researcher participated in the harvesting work, at least for some hours, in order to compensate for the time loss of the farmer during the information collection using a questionnaire.

4.6 Informal talks

Informal talks likewise were used to confirm and complement information collected using other methods. Sometimes, when people are formally questioned, they tend to give answers which sieve information they think lead them to risk in one way or another. During data collection, informal talks were useful in generating more information which could not be obtained during formal discussions, possibly because the respondents regarded the talks as causal whereby personal details (e.g. names) were not requested. In these talks, for example, they revealed unofficial methods used for controlling the waterweeds such as use of petrol to set fire to burn the waterweeds within the lake. They also exposed how some village leaders partake or are implied in illegal practices such as illegal fishing, charcoal business etc.

5 Secondary data collection

5.1 Documents perusal at government offices

At the village, ward, and the district offices, filed documents were accessed, perused and the relevant information/text sorted and photocopied for further analysis. In some research sites, there was either no electricity or photocopying services. In such cases, a digital camera was used to record relevant documents which were later downloaded to a computer, printed out, and sorted, and analyzed based on the objectives of the study.

5.2 Internet search

Due to limited accessibility of information at the national (Ministry) level (key informants at this level provided limited time for discussion), complementary information on institutional arrangements such as policies, acts, and strategies, were collected from the Internet by accessing websites of the relevant ministries (agriculture, livestock and fisheries). Other information such as speeches from the ministries was accessed in this way. Websites that proved useful in this regard include MIFUGO (www.mifugo.go.tz) and KILIMO (www.kilimo.go.tz). Other published documents were also accessed through the international Ramsar website (www.ramsar.org). In addition, a literature study of the existing research reports and some official statistics was conducted.

6 Data analysis

Two main methods were used for data analysis. The Statistical Package for Social Science (SPSS), and content analytical methods. SPSS was used for analysing the data collected through the household questionnaire. The data were coded, entered into a computer and then analysed. Descriptive statistics were calculated from the numeric data (e.g. the number of livestock, amount of crops harvested, etc.). Qualitative information collected from focus groups discussions, interviews, documents collected from different governmental levels, and from the Internet, were categorized into meaningful units and themes in keeping with the research questions.

Chapter 3.
Historical and administrative overview of natural resource management in Tanzania

3.1 Introduction

Before presenting the three case studies that are the focus of this work, this chapter outlines the historical background of natural resource management in Tanzania. This background covers changes in arrangements, institutions and actors over time. This historical background is important, as it provides an overview of the evolution of co-management between the government and the community in Tanzania. With such a picture in mind, we can comprehend how institutional changes have influenced the current natural resource management framework. An analysis of the current natural resource management administrative structure will follow this historical background. Natural resource management in Tanzania is currently implemented through a decentralised governmental framework. The role of the central government in this framework is to design national policy and legal frameworks, provide financial support to enable the participation of local governments and communities in implementation and enforce the policy and legal-institutional arrangements in natural resource management. This review of the current administrative structure will highlight the roles of institutions that link the central government to local governments in this decentralised framework.

3.2 Natural resource management eras and approaches in Tanzania

Natural resource management in Tanzania can be categorised into three eras: pre-colonial, colonial and post-colonial. Four natural resource management approaches emerged in these eras; namely, clan based, centralised (based in colonial, as well as post-colonial *Ujamaa*, policy), market based, and participatory natural resource management. The pre-colonial era was dominated by the clan based management regime, whereas during the colonial era, formal centralised natural resource management approaches became dominant. The centralised natural resource management approach dominated the early part of the post-colonial period, while later, market and participatory natural resource management approaches prevailed. Because it is impractical to isolate the management approaches from the eras (as the two overlap), in the following section they will be highlighted in an integrated way.

3.2.1 Pre-colonial natural resource use systems

During the pre-colonial era in Tanzania, there were no formal national policies to control and coordinate natural resource management. Informal arrangements and institutions oriented towards chiefdom systems governed the use of natural resources. Each ethnic group had a chief who oversaw the use of natural resources (Maghimbi, 1994). The chiefs enforced compliance to informal regulations and beliefs and imposed sanctions for non-compliance (Owino, 1999).

Elders assisted, and were answerable to the chiefs as the heads of specific clans, and their role was to oversee natural resources use in line with the traditions of their clans (Kauzeni, Shechambo, & Juma, 1998; Madulu 2005). These traditions included rituals, taboos and beliefs pertaining to the use of natural resources. The elders sanctioned those who broke the traditions. For example, for the *Pare* people of Mwanga, district fines could be paid in kind: a local brew and/or livestock could be confiscated from a violator regardless of his/her willingness to pay. Natural resource arrangements during pre-colonial times were thus in the form of top-down arrangements whereby a few individuals, i.e. the chiefs and appointed elders of clans, families, and tribes, enforced beliefs and traditional institutions while others in these communities were required to comply with the institutions (Kirk, 1999).

Traditional natural management systems had their own methods of avoiding conflicts in resource use. It was common to find multiple users of a natural resource of different cultural origins and with different interests in the same social ecological environment, such as livestock keepers and crops cultivators. In such situations, conflicts were unavoidable. Mechanisms to resolve conflicts among the users were therefore imperative. These included occupying different spatial positions within or between some ecological systems. While crop cultivators occupied upland areas (as these areas had reliable rainfall), herders occupied the lowland areas. An example of this was the mutual use of land resources by two ethnic groups, the *Maasai*, and *Chagga* on the slopes and lowland areas of Mountain Kilimanjaro. The advent of colonialism, nonetheless, interfered with these arrangements through the establishment of overlapping resource use arrangements where farming was practised, even in areas initially used for livestock grazing (Campbell, Misana, & Elson, 2004).

3.2.2 Colonial era

Formal relationships between the government and communities for natural resource management were formed during the colonial period. During this era, from 1888-1960, changes in the management of natural resources were introduced in the form of the exploitation of natural resources through regulations against traditional use practises. This era marked the beginning of state-led management regimes. The colonial government generally changed natural resource ownership arrangements from communal (clan based) to state based (nationalisation). A central change to the law declared the crown (governor) the custodian of the land. Emphasis was largely placed on cash crop production, and the allocation of land reflected this priority (Sundet, 2006). The government viewed local institutions and practises as threats to sustainable natural resource management. Clan based management institutions gave way to state-based management institutions, and government sectoral organisations were established to enforce these governing institutions (Owino, 1999; Van der Knaap, Ntiba, & Cowx, 2002).

The colonial government controlled the resources and the management arrangements. This eroded the power held by the elders in the communities prior to the emergence of state-led initiatives. The communities located near natural resources were forced to abide by these regulations through the establishment of punitive measures (Meroka, 2006). However, although the colonial government nationalised ownership of natural resources, traditional institutions retained some power over natural resources through chiefdoms. The colonial authority used an indirect ruling system whereby chiefs were used for collecting fines, and enforcing sanctions to non-compliance with the established regulations. Through this traditional governance, traditional institutions could still operate in some resource areas, though the custodian of the resources was the colonial government, who had the power to reallocate traditionally occupied land (Kauzeni *et al.*, 1998; World Bank, 2007).

Conflicts among resource users were inevitable during the colonial period. Based on their policy environment and objectives, the colonial government favoured crop farming. The farmers were, therefore, integrated with the colonial economy for the production of cash crops. Other livelihood practises, such as livestock grazing, were less favoured. This led to the alienation of grazing land and the displacement of herders from areas with water resources, as these were turned over to crop farming (Campbell *et al.*, 2004). The herders lost dominance over wetter areas on the rangeland, as well valleys and swamp areas, which they used during dry and drought periods. This power was transferred to the farmers, who introduced crops farming on these wetter and swamp areas (Campbell, 1981). In some areas, the colonial government established a partnership with herders by claiming that the government planned to assist the herders to develop livestock water infrastructures and pastures. However, the government later reallocated the land used by these herders to farmers (Hodgson, 2000). This resulted in conflicts, not only between the colonial government and herders but also between herders and farmers (Campell *et al.*, 2004; Hodgson, 2000).

To summarise the foregoing, during the colonial administration, there were two main issues pertaining to natural resource management arrangements: the institutional shift and increased focus on the farming, rather than herding, sector. There was an institutional shift from a traditional resource control framework to a governmental resources control structure in the form of divide-and-conquer arrangements. While before the colonial intrusion informal institutions governed resource use through traditions and beliefs, the centralised colonial authority displaced the informal institutions. Formal colonial institutions now oversaw natural resource governance. Regarding sectoral focus, cash crop production was more emphasised than other sectors, such as livestock. Institutional arrangements under the colonial administration remained in Tanzania until 1961, when Tanzania gained independence. Although natural resources remained nationalised, as was the case during the colonial period, the independent government adopted the collectivisation arrangement called the *Ujamaa* Policy.

3.2.3 Ujamaa policy

The third era (1961 to early 1980s) was the post-independence, *Ujamaa* policy era. *Ujamaa* is a Swahili word meaning socialism, adopted by the independent Tanzanian government in 1961. It forced people to come together into planned villages and mobilised their labour collectively in agricultural production (Mueller, 1980; Muzaale, 1988). Though the natural resources remained nationalised as a legacy of the colonial government, the *Ujamaa* policy eroded the authority of chiefs and traditional leaders that was recognised under the colonial administration (Mniwasa & Shauri, 2001). The mandate over ownership, utilisation and management of natural resources shifted from traditional leaders and chiefs to politically-elected village leaders (Campbell *et al.*,

2004). Emphasis was placed on the consolidation of bureaucratic institutions of the government (Mniwasa & Shauri, 2001).

Decision-making was one-sided during the implementation of the *Ujamaa* policy. Although the local people resided in the newly planned areas, they were not consulted before relocation or given an opportunity to give their views as to the suitability of the new areas for economic activities (e.g. for farming, livestock keeping, etc.). Government bureaucrats oversaw the people's relocation. The local people had to comply with the government's decision. Failure to follow this mandate resulted in coercive relocation enforced by government bureaucrats (Christiansson, 1988).

The *Ujamaa* policy had social, environmental and economic consequences. Socially, it extended the erosion of traditional clan based resource management organisations that had already started with enactment of the African Chief Ordinance of 1953 (Sheridan, 2004).[1] In fact, interference with the clan-based governance of resource use implied the freedom to destroy natural resources. The appointed government agencies were unable to oversee management of the vast natural resources in the country in a way that was efficient or effective given the personnel resources for this work. The end of colonialism thus not only marked the beginning of political freedom but also the undermining of the environmental conservation initiatives and legislation initiated during the colonial periods (German, 2004; Kauzeni *et al.*, 1998). Resources that had been successfully protected under the traditional common property arrangements became nobody's resource with the repeal of chiefdom systems, and experienced open access exploitation. Environmental degradation was therefore an inevitable consequence (Meroka, 2006; Sheridan, 2004). Plans drawn by the state that disregard existing complexities of local social and ecological systems certainly result in instability of these interdependent social and ecological systems (Scott, 1998).

Ujamaa policy paid little attention to natural resource conservation but strongly emphasised economic growth. The collective strategies were implemented in a top-down style. The local people had to receive and implement commands to move to planned villages without being consulted as to whether those areas were suitable for economic activities (Christiansson, 1988). Later, the appropriation of private natural resources for economic development was replaced by the state's engagement in large-scale production through its parastatals. However, contrary to the government's mission to boost the economy through the *Ujamaa* policy (Kauzeni *et al.*, 1998), the economy significantly deteriorated (Meroka, 2004). The investment in government driven-projects did not yield the anticipated results, largely due to poor management and corruption (Assens and Jensen, 2003), poor law enforcement, poor policy implementation, and the lack of adequate skilled workers to run the projects (Limbu & Mashindano, 2002). The parastatals became both inefficient and ineffective, leading to economic deterioration (Assens & Jensen, 2003; Barrow & Mlenge, 2003). Thus, both natural resource degradation and economic deterioration were the result.

Furthermore, the *Ujamaa* policy put more emphasis on some sectors than others. It favoured crop production, relocated people into collective settlements, and established collective crop farms. For example, a land reform policy was introduced and implemented that reallocated most

---

[1] The African chief ordinance of 1953 repealed the authority of the chiefs. Because after independence the trend was towards nationalistic unity, the party (CCM) bureaucrats negatively regarded the chiefs as having power in their clans that could be used to divide the people into tribal based fragmentations, harming state nationalism.

of the areas that were used for grazing to crop farming under collective production arrangements. Because of this, herders were deprived of grazing resources, ultimately leading to conflicts between herders and farmers (Shivji, 1997). Sectoral conflicts that prevailed during the colonial administration era escalated during the *Ujamaa* policy tenure. For example, the enforcement of land redistribution reform arrangements partitioned land into small administrative units (villages) for collective agricultural production, a shift from common property and a family-based economy to collective production, adversely affecting the livestock keepers. It was estimated that because of this institutional reform, about 250,000 livestock keepers lost their land and had only restricted access to pastures and wells. The livestock keepers could hardly exercise the migratory grazing practises they were accustomed to because sedentarisation was viewed as a prerequisite for modernisation (Kirk, 1999).

The *Ujamaa* policy did not accomplish its intended goals in Tanzania. The top-down interactions between the government and the local people in natural resource management during the *Ujamaa* era was unsuccessful in improving either economic growth or conservation of natural resources. Social, economic, and environmental deterioration were the inevitable consequences. An alternative approach was required to improve these situations. The government of Tanzania launched internal economic recovery programmes to stabilise the deteriorating economy. These included the National Economic Survival Programme (NESP), which aimed to increase production and exports, as well the structural adjustment programme (1982-1985), which aimed at pruning the central government budget and restructuring public enterprises (Meertens, 2000). These reform and stabilisation strategies did not work. At the same time, the World Bank and IMF placed pressure on the government of Tanzania to deregulate all spheres of its economy as part of its reform program; otherwise, the country would not receive financial assistance from these institutions (Meertens, 2000). Although the government initially resisted the adoption of the IMF and World Bank programmes because they were not in line with their socialist strategy, later, as the internal reform programme failed, it was forced to comply (Limbu & Mashindano, 2002). The country thus turned to market-based approaches.

3.2.4 Market economy era

The failure of *Ujamaa* policy, with its consequences for economic, social, and environmental deterioration, forced Tanzania to adopt market-economy policies in the mid-1980s. In this era, the government's control over the economy diminished, while the influence of market forces increased. Market liberalisation led to the integration of local resources with large market systems at the regional, national and international levels. This resulted in the escalation of over-exploitation and the degradation of natural resources in some areas, such as the southern highlands of Tanzania. In these highlands, the commercialisation of agricultural production resulted in the transformation of farming practises, whereby farmers came to cultivate fast-growing crops using unsustainable farming practises to gain a cash income as rapidly as possible. Farmers shifted from cultivating pyrethrum and green peas using ridges and terraces to cultivating wheat and Irish potatoes in a flat cultivation system. This shift resulted in soil erosion (Sokoni, 2001).

Whereas economic liberalisation can be commended for reducing or eliminating situations wherein one individual or group of individuals benefit from actions and efforts of others in natural resource management, especially when common property is converted to private management regimes, Tanzania has also suffered some negative results. Because most land resources are communal, not common, liberalisation led to increased inequality in natural resources access. For example, some groups in the community are now able to sell communal property, thereby benefiting at the expense of others. Other local people cannot manage their natural resources in sustainable ways because, among other reasons, they lack adequate capital and labour in light of agricultural commercialisation. Large farm plots in particular have difficulty maintaining sustainable farming practises such as the use of ridges and terraces, and face poor markets for agricultural produce. Further the change of cropping systems to favour short-term crops is destructive to contour bunds and terraces regarded as fertile areas for cultivating crops (Sokoni, 2008).

The transition to a market economy has transformed both traditional and governmental institutions and arrangements. Commercialisation has altered traditional land use and allocation arrangements. Traditional methods of allocation and inheritance of land have given way to new arrangements, wherein land is sold or leased to individual farmers even outside a clan. This negatively affects the availability of land for members of a clan, and stands in contrast to earlier eras, when elders enforced land allocation arrangements and made sure that all members of the clan were given some land. On the side of the government, commercialisation has led to the sale to private ventures of village land that had previously been available for the use of villagers as a whole. This is supported through the land policy of 1995, which officially allows individuals to obtain land for agricultural investment provided this land is not under communal use, used by community projects in the village, or a conservation area. This resulted in competition for land and natural resources between large-scale farmers and small-scale farmers and between pastoralists and crop cultivators, among other interests. It is rare today to find unoccupied land in many villages in Tanzania (Sokoni, 2008). Market based policy has therefore resulted in an institutional shift from clan-based and socialised arrangements to individual or private institutional arrangements, and have impacted access, use and management of land resources.

Market liberalisation contributes to conflicts among resource users in Tanzania. Transformation of natural resources under traditional institutional use arrangements (i.e. under the clans) or under a group of users (e.g. communal grazing land) to private management arrangements creates tension and insecurity among resource users. The commoditisation of agriculture and the encouragement of private tourist industry, for example, have led to further alienation of grazing land previously under communal management. This has led to conflicts between private investors and herders (Campbell *et al.*, 2004). In some areas, even the government has been involved in land sales, whereas at the village level, land used by herders is appropriated and sold by the village government to large-scale farmers. This has escalated conflicts, not only between users of communally owned land and private investors but also between users of the communal land and the government (Rie, 2002). In these situations, traditional institutional arrangements of resource use based on social solidarity among users are undermined and replaced by individual arrangements that favour individuals with command over resources. As a result, private resource use arrangements unbalance resource accessibility among individuals in the community. Few can access the resources, and the majority has limited or no accessibility. The limited accessibility of the resources may lead to overuse pressure and the degradation of the limited resources that can be accessed (Sokoni, 2001). Under such situations, environmental conservation partners and stakeholders may recommend participatory natural resource management to alleviate unsustainable natural resource management.

3.2.5 Participatory approaches

Community-based natural resource management emerged in Tanzania in the 1980s. Its inception was due to recommendations by donors and conservation agencies that the government had to adopt this approach as it had failed to implement effective natural resource conservation policies (Songorwa, 1999). Its adoption was not universal, however; community-based natural resource management varied in the time of its inception depending on the resource. For example, Willy (2000c) mentions that the first participatory community based forests were established in 1994, while Songorwa (1999) notes that the community-based management of wildlife resources began earlier. Though there are some discrepancies in the literature regarding the inception date depending on the resource under study, we can generally state that community-based natural resource management began in the late 1980s.

The emergence of participatory natural resource management was expected to resolve various issues in the management of natural resources, including resource ownership. Under this management system, integration and the consideration of rights of resource users is viewed as a method of encouraging active participation in resource management. Studies of resource management in Tanzania show that when the local people see themselves as the owners and co-managers of natural resources in their vicinity, there is a sense of concern about using a resource in sustainable ways. Similarly, a feeling of commitment to monitoring the use of the natural resources by external users has been observed (Willy, 2000c). The provision of secure ownership rights and the authority granted to the community over management of natural resources may make community members accountable in implementing management plans and enforcing rules to limit the unsustainable use of natural resources (Odera, 2004).

Participatory natural resource management in Tanzania influences the contentious relationships among resource users, though information on the nature of the influence is mixed. In some cases, the establishment of participatory natural resource management has succeeded in improving social relationships between resource users. This has occurred through the formation of representative conflict resolution committees integrating formal and informal institutions in the villages, and the establishment of mechanisms to punish non-compliance with sustainable natural resource use and management (Kajembe, Monela, & Mvena, 2003). Other experiences, however, show that the establishment of participatory management may transform existing relationships, not only within the community but also between the community and external users of a particular resource. Some contentious relationships may continue to develop after the implementation of participatory management as a result of external and internal resources, user groups who depend on and fight for the same resource for food and income, and heterogeneity within the same community the result in conflicting interests, purposes, and marginalisation of the interests of some members of the community. These conflicts render the potential for natural resource sustainability questionable. At other extremes, government agencies with different interests may conflict in their interactions with the community. While the interests of one of the agencies may be congruent with the interests and objectives of the community, the objectives of another agency may contradict those of the community (Nelsen, 2004).

Leadership may also influence the participatory management of natural resources. Leadership involves leaders from the community level to the governmental level. Leadership can be independent or can combine traditional or formal powers. The record of participatory natural resource management in Tanzania indicates that when leadership is capable, stable, and accountable there is higher chance of sustaining collective natural resource management. In contrast, when the leadership is unstable and/or unaccountable, it is likely that conflicts between parties will dominate participatory natural resource management, which may lead to its ultimate failure (Nelsen, 2004). While Nielsen (2004) notes that an increase in revenues and employment for villages in Northern Tanzania results from participatory natural resource management, he also notes that conflicts emerge alongside these benefits. Such conflicts can be attributed to unstable governance, poor decision-making and poor management of revenue. Although village leaders usually serve a term of five years before a new election, the community in one village (Sinya village) changed village leaders three times within one five-year term because the leaders misappropriated the revenues that from the community's participation in eco-tourism and environmental management projects. This is indicates that although community participation in natural resource management may have the potential to generate both socioeconomic and environmental benefits, other factors within the community may constrain the sustainable realisation of these benefits and hence discourage community participation. In contrast to this negative result, however, Baldus, Bensen and Siege (2003) find that in Selous National Park, good leadership was an important dimension that enabled the community to organise for participatory conservation work, and to share in the benefits accrued from their efforts. To manage revenues accrued from their participation in wildlife management, the community and the village government established a village committee for overseeing management arrangements. The sharing of the benefits of this participatory initiative is discussed among all villagers in the village assembly in an open and democratic way, and profits from the management work are invested in other community projects or saved in a community bank account. These contrasting findings indicate that, depending on how the leadership is structured, it may contribute to or constrain sustainable participatory natural resource management.

3.3 Current administrative natural resource management structure

So far, this chapter has highlighted the evolution of natural resource management in Tanzania. It has covered several approaches: namely, clan based, top-down (colonial and Ujamaa approaches), market based, and participatory natural resource management regimes. However, there has also been a change in the administrative structure from a centralised to a decentralised governmental structure that has accompanied the shifts in management approaches. The natural resource management regimes in the present era are implemented under a decentralised governmental structure. In this section, the current administrative structure of natural resource management will be explored.

Currently, natural resource management in Tanzania is implemented within a decentralised framework (Figure 3.1). The argument in support of decentralisation of natural resources in Tanzania is that when the central government had previously concentrated its power and authority

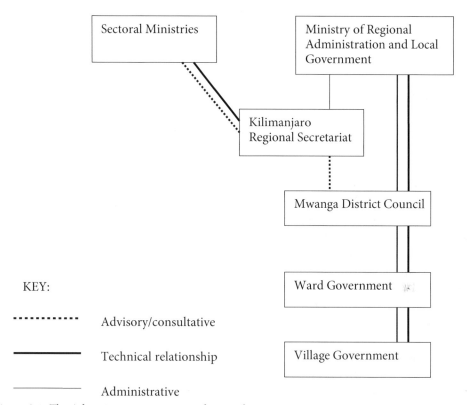

Figure 3.1. The Administrative structure of natural resource management in Tanzania.

over natural resource management, it could not provide practical protections against natural resource degradation (Mniwasa & Shauri, 2001). This is an extension of the understanding that the degradation of natural resources was caused by the failure of the top-down natural resource management approach. However, this extended approach recognises that the failure of top-down natural resource management regimes does not fully explain natural resource and environmental degradation in Tanzania. Other factors, including the concentration of the decision-making and ownership powers by the central government, also contributed to environmental degradation. In response to these problems, the central government has since granted the responsibility of enforcing sustainable management of natural resources to local government authorities. The Local Government Act No. 6 of 1999, which amends the Local Government Act of 1982, argues that because local governmental institutions are situated in local areas, they are in a better position to deal with environmental and natural resource management issues than the central government, which is situated further from the sources of the issues. In turn, the central government is required to empower local governments and community actors to properly engage in resource management in their entrusted jurisdictions.

At Lake Jipe, the three sectors under study (agriculture, livestock and fisheries) present some organisational differences in natural resource management. While for some sectors formal organization of local resources users exists for others informal organization exists. In the fisheries sector, village-level committees established under national fisheries policy and law represent fishing communities in the management of fisheries resources. In other sectors, there are no formal sectoral committees for the management of natural resources. Instead, there are informal committees formed by the livestock keepers and the farmers themselves. In most cases, these are designed to defend users' interests. For example, the livestock keepers' informal committee was formed for the purpose of protecting pastoral land against encroachment from other land users. Similarly, farmers formed informal committees for negotiating and sharing water resources among themselves. Whereas the formation of the fisheries committee was influenced by formal governmental institutional arrangements, the formation of the livestock keepers' and farmers' committees evolved in the process of resource use to mediate between the users. Besides these formal and informal committees, each sector is represented in a formal multidisciplinary environmental management committee that governs the management of natural resources in the villages as a whole.

In the coming sections, I briefly highlight the relevant actors in the implementation of natural resource management within the framework of decentralised local government. These actors include Sectoral Ministries, the Ministry of Regional Administration and the Local Government, the Regional Secretariat at the regional level, and the district council. Under the district council, in line with the current study, there are Departments of Natural Resources in which fisheries are integrated, and the Department of Agricultural and Livestock Development. This departmental separation continues along the administrative levels at the wards, division, and villages.

3.4 Ministerial level

Sectoral ministries represent one level of administrative structure in natural resource management in Tanzania. Two of the sectoral ministries are examined in this section: the Ministry of Agriculture, Food Security and Cooperatives, and the Ministry of Livestock Development and Fisheries. Under the current managerial framework, the central government, and therefore the ministries as well, is charged with (among other things) preparing national sectoral policy environments. The following questions will be answered in this section: how do the sectoral ministries relate? How do the sectoral policies formulated by these ministries involve the local community at their preparatory stage, and how do these policy instruments accommodate community participation in management of the natural resources used in the livestock production, agricultural production, and fisheries management?

3.4.1 The relationship among ministries

Relationships among the three sectors at the ministerial level are complicated. While some sectors fell under one ministry at one time and then separated to two different ministries at another time, others have undergone the opposite transition. Until the late 1990s, the agricultural and livestock sectors fell under the same ministry, the Ministry of Agriculture and Livestock Development. At the time of writing, the two sectors fall under independent ministries. These changes have not only occurred structurally but also in policy planning. While before the separation of these ministries, and some years afterwards, the livestock and agricultural sectors had were integrated in policy planning, since 2006 the two sectors have had independent policy planning processes. While the agricultural and livestock sectors came to be separated into separate sectors, the fisheries and livestock sectors underwent the opposite process, integrating to form a single Ministry of Livestock Development and Fisheries.

The changes in integration among the three sectors can be explained by a combination of inter-sectoral conflicts and politics. Agricultural and livestock staff and professionals at the ministerial level hold opposing points of view regarding the relationship between the two sectors. Agricultural professionals and staff argue that the two sectors should fall under one umbrella ministry because the majority of rural people in Tanzania are agro-pastoralists; therefore, to address the challenges these clients face, management and planning for the two sectors must be centrally integrated (MAC, 1997). Their counterparts specialising in livestock management claim that, under the integrated ministries and policy planning arrangements, the livestock sector has been given less emphasis (MLD, 2006). These stakeholders favour the view that for the development of the livestock industry requires an independent Livestock Ministry responsible for central livestock planning.

The integration of the fisheries sector into the Ministry of Livestock Development (MLD) resulted from political influence. Whereas the fisheries sector before its integration in the MLD was a division of the Ministry of Natural Resources and Tourism (MNRT), following Tanzanian cabinet reshuffle in the early 2008, this sector was removed from the MNRT and placed into the MLD to form the Ministry of Livestock Development and Fisheries (MLDF). The reasons for this decision, according to President Jakaya Kikwete of Tanzania, is that the fisheries sector has been underemphasised because the MNRT is laden with other vast sectors of forest and wildlife management. The President asserted that livestock and fisheries sectors are related and that fisheries could be more productive if given a higher priority. These arguments may seem convincing but it remains to be seen whether the placement of fisheries sector in the MLDF will, in itself, result in improved fisheries productivity, and whether it will lead to an effective use of the fisheries' existing potential. Other factors, such as good leadership, good planning, accountability, etc. can influence the development of the sector regardless of which ministry provides oversight.

Although at the central level the livestock and agricultural sectors are separated, at the local level the two sectors are integrated into one department. It is possible that integration of centrally separated sectors at the local/district level will give the district department an opportunity to mediate contradicting and uncoordinated institutional arrangements from the central authorities. However, it is also possible that this difference at the local level will result in confusion within the district departments, especially when there is a limited capacity to translate and understand the institutions from uncoordinated central levels. This latter possibility seems less likely, though, because the district departments are under the authority of the district councils that are responsible to the Ministry of Regional Administration and Local Government (MRALG) that coordinates local government administration and central sectoral policies and laws. The MRALG was established in 1998 to coordinate sectoral institutions, policies and programmes from the line ministries, and to oversee the implementation of these institutions in the local areas. In essence, the MRALG has technical and administrative staff that coordinate and communicate with the local government authorities where policies and programmes are to be implemented (Baker, Wallewik, Obama & Sola, 2002).

The establishment of the MRALG implies the some problems with direct communication between the sectoral ministries and their staff at the district levels (CARE 2006). However, there is duplication of chain of commands (Dallu, 2002). The local government staff receives commands from the sectoral ministries, and from the ministry overseer of decentralisation, i.e. MRALG. If the commands and institutions from MRALG and sectoral ministries are poorly coordinated, the result is a conflicting implementation of the institutions and programs at the local level. Proper coordination of the institutional arrangements of the MRALG and sectoral ministries is, therefore, imperative to avoid potential contradictions in the implementation process at local level. The question, nonetheless, is whether the MRALG has the power to manage these challenges and positively influence the management of natural resources.

The formation of the MRALG has resulted in the sharing of authority and roles between the central sectoral ministries and MRALG. The MRALG is now the institution responsible for employing the staff that run the sectoral institutions at the local level. While sectoral ministries used to be responsible for employment, the staff are now answerable to the MRALG. Therefore, while the sectoral ministries have the jurisdiction of formulating the policies and laws for the management of natural resources at the local levels, the MRALG has the jurisdiction of coordinating the sectoral institutions and their implementation at local level, and of employing the technical staff who facilitate and oversee the implementation and enforcement of the sectoral institutions. Thus, in essence, and with exception of the advisory communications (done through representatives of the sectoral ministries at regional level), the institutional relationships between the sectoral ministries and sectoral departments at the local level are mediated by the MRALG.

Because the institutional arrangements of the three sectors, though prepared at the central level, are implemented at the local level, conflicts and contradictions among the institutions and actors at the central governmental level may negatively affect the implementation of these arrangements at the local level. The conflictive and contradictory sectoral institutions may disrupt each other's governing potential of one another, making it difficult for local actors to implement their policies. This may also make it difficult for sectoral actors at the district level to design the supporting institutions and laws for the enforcement of the mandates of the national sectoral institutions. In this case, it may be difficult to achieve collaborative planning and to navigate the competing influences of the different sectors. In turn, this may impair the collaborative and sustainable management of natural resources for agricultural, fisheries and livestock production at the Lake Jipe social-ecological environment.

3.4.2 Participation of local community in policy processes and natural resource management

For all three sectoral policies, the local community is involved in the policy process through a consultative approach. The process is overridden, however, by relevant central governmental agencies (agriculture, livestock and fisheries) that, based on social, political, and economic pressures (local and global), prepare agendas for policy formulation or revision. The users at the community level can be represented through the users associations or through relevant local governmental departments. In addition, in zonal meetings that usually take place each quarter, the local authority representatives participate in discussing various issues regarding natural resource management. Deliberations from these meetings are transmitted to the relevant sectoral departments at the central governmental level.

Participation of the local community in the conservation and management of natural resources is explicit in fisheries policy but implicit and unclear in livestock and agricultural policies. The fisheries policy of 1997 clearly states that for sustainable fisheries management, communities in the vicinity of fisheries resources must participate in fisheries management through the formation of fisheries committees appointed by fishing communities. In contrast, livestock and agricultural policies (Agricultural Policy of 1997, and Livestock Policy of 2006) emphasise the commercialisation of agricultural and livestock production, and claim that the integrated and sustainable management of natural resources that these production activities require will be promoted along with the livelihood objectives. However, the role of local communities in the management of natural resources in these transformations is unclear.

3.4.3 The regional secretariat

The regional governments facilitate sectoral policy and plan implementation at the local governmental level. A Regional Secretariat (RS) links the central government and the local governmental authorities. The RS is comprised of the Regional Commissioner (RC), the Regional Administrative Secretary (RAS) and four technical clusters; namely, management, economic and development, social services, and physical planning and engineering. The RS has the role of facilitating the implementation of sectoral policies. Included in the economic and development technical cluster are technical staff, including agricultural, livestock, and fisheries officers. They form one of the four technical clusters of the RS. They are employed by the sectoral ministries but are attached at the regional level. Before decentralisation, the role of these technical staff was to implement the sectoral policies. Within the current decentralised structure, their role has changed to facilitators. They facilitate the sectoral departments at the local government (district) level in their implementation of policies by providing them with expertise and capacity-building programmes (Baker *et al.*, 2002). Technical personnel at the district level have assumed the responsibility of implementing national sectoral policies that once was implemented at the regional level.

Although the RS is regarded as a linchpin linking the central government and local governments, in reality it is constrained in performing this role. For example, in the Kilimanjaro region where this study was conducted, technical members of the RS have limited facilities, are poorly informed, and have limited financial resources to enable them to enact their roles effectively. The technical officers at the local governmental level (the district level) are better informed than the technical officers at the regional level. In other words, there is evidence for poor management by the technical officers at the regional level. In part, this occurs because communication between the sectoral ministries and the local government authority bypasses the RS. In addition, training may be given to the officers at the local governmental level by officers from the sectoral ministries without an awareness of the officers at the regional level. The RS, therefore, is both poorly informed and lacking in the necessary training. Furthermore, due to financial limitations, the RS technical officer sometimes has to depend on local governmental authorities for transportation to the field to assess the implementation of the sectoral development arrangements. All of these constraints make the technical officers at the RS level less confident and capable of implementing their responsibilities effectively (Regional Agricultural Advisor (RAA), Kilimanjaro Region 2008).

Due to a lack of technical and physical resources, the government is therefore limited it its ability to enforce sustainable natural resource management on its own. This problem calls for co-management arrangements where the government entrusts some of the management to the local people, and collaborates with informal institutions at community level, which may reduce costs to the government while enabling sustainable enforcement of sound management practises.

3.4.4 Local /district government authority

There are three important institutions at the local governmental level: the district council, the ward development committee, and the village council. The agricultural, livestock and fisheries sectors are represented at these councils. At the district level, there are two departments relevant to this study thesis: the Agricultural and Livestock Department, and the Natural Resources Department, which includes the fisheries sector. As introduced in 3.3.1, in contrast to the ministry level, the livestock and agricultural sectors are integrated at the district level as the Agricultural and Livestock Department. In addition, while the livestock and fisheries sectors are integrated (part of one ministry) at the central governmental level, at the district council level they are separated; the fisheries sector falls under the Natural Resources Department.

The district-level departments existing are entrusted with implementing the agricultural, livestock and fisheries programmes and policies. They are responsible for drafting programmes, plan of actions, strategic plans, by-laws, etc., in keeping with national policy and its legal instruments. In other words, they have the responsibility of adapting national institutional arrangements (policies, laws, Acts) to local situations. This level is also responsible for coordinating participatory arrangements for natural resource management and the livelihood demands of the local people. To use the example of the fisheries, at this level, participatory planning and deliberations are undertaken on how to involve the local fishing community in participatory management of fisheries resources in keeping with the national fisheries policy.

Ideally, communication between the sectoral ministries and departments at the district council level has to be coordinated through the Ministry of Regional Administration and Local Government. However, in practise, not all communications follow this path. The sectoral ministries and the Ministry of Regional Administration and Local Government may have parallel communication with the district sectoral departments at the district level and this can cause contradictions, conflicts, and dilemmas for actors at the local governmental level.

At the ward level, there are representatives of the agricultural, livestock and fisheries sectors, although they are not equally available at every ward. Besides the technical people at these levels, there are administrators who oversee all executive aspects for villages under their jurisdictions including agriculture, livestock, fisheries, and the use of natural resources in general. At this level, there are sometimes not adequate staff to act as technicians, so administrators have to fill the technical role.

The village is the closest level to the users of natural resources in fisheries management, agricultural production, and livestock production. Administrative and technical roles are performed. However, at this level too, there is generally a shortage of technicians, and therefore the administrative officers sometimes fill their positions.

3.5 Conclusion

This chapter briefly described the administrative structure in the implementation of natural resource management in Tanzania. In brief, it has depicted the roles of relevant government institutions along with the challenges and constraints to effective management that accompany those roles. This chapter has also mapped out the history of resource use. It has uncovered the path dependent sectoral approach in the management of natural resources that led to the current government structure. It further indicates that the failure of the government to manage natural resources in a data-poor environment has led to some management jurisdictions being entrusted back to the local people. In other words, management is returning to the informal/customary institutions ignored in the colonial, Ujamaa and market periods.

In the next three coming chapters, empirical cases of co-management of natural resources between the government and the local people at Jipe Lake in Mwanga district, Kilimanjaro region in Tanzania will be discussed. These three empirical chapters will investigate collaborative arrangements in natural resource management between the government and the resource users at Lake Jipe. The investigation entails the analysis of interactions of the government and the community, in among other aspects, resource management, institutional enforcement, and problem resolution. Because there are multiple diverse formal and informal institutions that govern and mediate the relationships among the stakeholders in natural resources for the three sectors, dynamic interactions exist within and between these institutions, as well as with governmental and community actors. These interactions have implications for the use of these natural resources because their nature determines the sustainability of use of the natural resources in the Jipe social-ecological system. Sustainable management of natural resources at Lake Jipe, therefore, depends on interactions between formal and informal institutions, practises based on actors' objectives, and the effects of these interactions on natural resources systems and the social-ecological system as a whole.

Chapter 4.
Management of fisheries resources at the Lake Jipe wetland in Tanzania

4.1 Introduction

Fisheries resources at Lake Jipe bring together various government and community actors. The Natural Resources Department at Mwanga district is entrusted with enforcing and overseeing management of fisheries resources at the district level, while the community implements policies and by-laws through fisheries committees. The local fishing community comprises two geographical categories of fishers: residents and non-residents. The category of resident fishers includes fishers that come from an upstream area of Lake Jipe. Non-resident fishers generally originate from the southern regions of Tanzania and nearby regions, such as Tanga, Arusha and Kilimanjaro (Figure 4.1). Migratory fishers move from one place to another based on the availability of fish.

The Mwanga district has adopted a national approach which emphasises collaboration of local governments and local people in the enforcement of fisheries institutional arrangements. At Lake Jipe, a fisheries committee has been formed to guide the enforcement of collaborative fisheries resource management. The committee was established in keeping with the institutional

Figure 4.1. Regions in Tanzania from which fishers and migrants to lake Jipe originated.

requirements stipulated in the fisheries policy of 1997 and the Fisheries Act 22 of 2003. The current dominant assumption is that institution of participatory management is a solution to unsustainable management practises. Using the case of Lake Jipe, this chapter investigates the collaboration between the government and the community in the management of fisheries resources at Lake Jipe. Specifically, the chapter first analyses fisheries management practises and their impacts on fisheries resources in the Lake Jipe area. Second, it examines how the government and community collaborate in managing sustainable fisheries practises. Third, it explores the participation of government and local institutions in resolving conflicts that emerge in fisheries management. Finally, it analyses and evaluates co-management arrangements that exist in fisheries management in the Lake Jipe area.

4.2 Background on fisheries at Lake Jipe

This section highlights four aspects of fisheries resource management at Lake Jipe: the evolution of fisheries activities, the status of fisheries resources, the use of fisheries resources, and actors involved in fisheries resource management.

4.2.1 Evolution of fisheries activities at Lake Jipe

Fishing has been one of the main livelihood activities conducted at Lake Jipe since the 1950s. By 1964, there were 150 fishers at eight camps along Lake Jipe namely Mkisha, Jipe, Ubungu, Urondo, Makuyuni, and Mirangeni. Some of these camps are now villages, while others vanished as people moved out. Many of the fishers in the area were newcomers from other countries and regions, including people of groups such as the Jaluo, the Kisi from Kenya, they Nyasa, and the Manda from southern regions of Tanzania. These newcomers came to mix with the native *Pare* people. After 1968, the number of fishers increased to 2,000, and in the same period the Nyumba ya Mungu dam was established. Some fishers moved to this dam where they established new settlements that still exist today. These settlements are Kagongo, Bora, Nyabinda, Njia Panda, Handeni, and Kiti cha Mungu (Mwanga, 2006).

According to the elders at Lake Jipe, fishing activities at the lake are generally believed to have been started by newcomers from Tanga, Songea, Mwanza, Mbeya, Kigoma and other regions of western and southern Tanzania who were originally brought to the Tanga region during colonialism as migrant workers to provide forced labour in sisal plantations. During the construction of the central railway infrastructure, these labourers were used to build the railway line from Tanga to Kilimanjaro, and it is during this time that they noticed the existence of Lake Jipe. Some returned to their home regions and reported the location of the lake to friends and family, while others moved to the Lake Jipe area directly after completing construction of the railway line. It is said that in the 1950s this important livelihood practise was mainly carried out by the resident fishers born in the vicinity of the lake. These early resident fishers fished at the subsistence level, and limited their fishing activity to inshore trapping using long sticks due to their fear of wildlife attacks from crocodiles, hippos, and lions. The arrival of newcomers from the southern and western regions of Tanzania marked the beginning of large scale fishing activities. Fishers from these regions began implementing commercial fishing practises, including the use of dragnets.

From the 1960s onward, the Lake Jipe region experienced the integration of locals and outsiders in fishing activities at the wetland.

As mentioned above, the arrival of non-resident fishers in Lake Jipe transformed fisheries practises. Elders claim that before this period, illegal fishing was not practised. This is because only large fish, of more than three to five inches, were captured at the time. Illegally harvested small fish would have been unmarketable due to the abundance of large fish. However, demographic changes introduced unsustainable fishing practises into the area. When non-resident fishers first arrived in the 1960s, according to elders at Lake Jipe, they killed many crocodiles, and so were able to engage in more offshore fishing. Over time, resident fishers came to engage in offshore fishing as well. This led to the decline in the fisheries industry of Lake Jipe that began in the 1970s.

Fishing is a dependable activity that contributes to the livelihood of local people. It is estimated that of all livelihood practises of the area, fisheries contribute approximately 60% of the total food security (Mwanga, 2006). There is, however, not much information regarding the contribution of Lake Jipe fishing to the national economy. Planners and economists who study this sector commonly estimate the contribution of the fisheries sector in Tanzania on the basis of three main inland lakes (namely, Lake Victoria, Lake Nyasa, and Lake Tanganyika) and to generalise the performance of these great lakes as the total contribution the economy of the fishing sector. The contribution of these big lakes to the national economy has been well understood. However, though overlooked, Lake Jipe and other small lakes of its kind are important sources of subsistence for inhabitants, and also provide employment opportunities to fishers, fishing communities, and non-residents engaged in fisheries transactions, thereby contributing to the district and regional economies.

4.2.2 Status of fisheries resources at Lake Jipe

The fish at Lake Jipe have been in decline from the 1970s to the 2000s (Table 4.1). It was not possible to get statistical data from the district and national fisheries authorities on changes in fisheries production at Lake Jipe. The information used in this study is based on anecdotal information from fishers at Lake Jipe based on their experiences and memories of changes in catches per unit efforts for the past three decades. This decline is depicted in Table 4.1. In the 1970s, the number of fish per catch ranged between 3,500 and 4,000, in the 1980s it ranged between 2,000 and 3,000, in the 1990s the number ranged from 900 to 1000, and for the 2000s it ranged between 300 and 700. Not only has the amount of fish declined from 1970s to the 2000s, but the fish size has also decreased. Whereas in the 1970s and 1980s large fish of three to five inch sizes were generally captured, in the current years, fish of sizes less than two and half inches are more common in each catch (Table 4.1).

Fishers and fisheries authorities disagree as to the primary cause for the decline in fish availability. Fishers at Lake Jipe argue that the proliferation of water-weeds, which increased in the 1970s, has given the fish places to hide, making them difficult to access. Fishers support this argument by noting that when they clear some areas of water-weeds, they are able to catch larger fish. This indicates to them that the best solution to fish unavailability is to clear the water-weeds.

Although the water-weeds are a problem, fisheries authorities note that unsustainable fishing practises are a problem as well. By placing the entire explanation for low catch quantities and small

Table 4.1. Decrease in catch quantity and fish length per catch effort from the 1970s to the 2000s (source: fishers at Lake Jipe, 2009).

	Fish per catch	
Years/decades	Number of fish (range)	Frequent catch sizes (inch)
1970s	3,500-4,000	4-5
1980s	2,000-3,000	3-3.5
1990s	700-1000	2-3
2000s	300-700	<2.5

fish sizes on water-weeds, fishers justify their fishing practises, which in fact have a negative impact on fish availability at Lake Jipe. From an interview with fisheries officers at Mwanga district, it appeared that while they too view water-weeds a problem that threatens the sustainability of Lake Jipe, they note that the use of unsustainable fishing equipment plays an equal role in contributing to the unsustainability of the lake.

Fishers' view of water-weeds as the primary problem may be attributed to the fact that water-weeds reduce the fishing space available to set their nets. In principle, this seems a physical obstacle to their livelihood opportunities. It seems that their major concern is fishing space. The fishers are not reflexive on their own practises, however. Greater awareness may be needed to enlighten fishers on detriment to fisheries sustainability caused by fishers' unsustainable practises.

4.2.3 The use of fisheries resource

The use of fisheries resources can involve both good and bad practises. Co-management can benefit from good practises, and, can, in turn, enhance these practises' contributions to sustainable fisheries management. Co-management can also benefit fisheries sustainability by understanding bad practises and developing the strategies to change these practises to have a positive impact on the sustainability of fisheries resources. Because of this, it is important to develop a picture of both the benign and damaging fishing practises at Lake Jipe.

The harmful fishing practises at Lake Jipe involve the use of destructive fishing tools that degrade the lake ecology. These tools include nets whose wholes are smaller than the recommended two or more inches (according to fisheries by-laws at Mwanga) and mosquito nets, which degrade immature fish stock. This endangers the future of the supply of fish resources. According to established institutional arrangements (e.g. the Fisheries Policy of 1997), the use of destructive fishing equipment results in the exploitation of juvenile fish stock and, thereby negatively impacts the reproductive potential of future stock. The continued use of these destructive tools will result in a further decline in the availability of sizeable fish. In response to declines in the availability of fish, fishers become more likely to turn to illicit methods in their fishing to meet their livelihood and income needs. The use of illicit nets therefore results in a vicious cycle of fish stock degradation.

Other practises at Lake Jipe have positive implications. Some practises, though carried out at a small scale, have the potential to conserve the environment if they are enhanced. One such practise is the manual removal of water-weeds to create spaces so that fishing hooks, nets and fish traps may be set. The manual removal of water-weeds (by cutting) to create spaces for fishing indicates that fishers are capable of working together to achieve broad goals, which could be applied to conservation.

4.2.4 Actors involved in fisheries management at Lake Jipe

Fisheries resource management at Lake Jipe comprises multiple governmental and community actors. While some actors (e.g. the District Natural Resources Department) are entrusted with facilitatory and regulatory roles, others enforce, monitor, control and survey management practises (e.g. the fisheries committee) and mediate resource use disputes among resource users (e.g. the elders and the District Commissioner). The presence of multiple government and community actors indicates interplay of multiple and complex systems of norms, rules and regulations in the use of natural resources. The existence of multiple actors may imply the existence of tension among their values, capacities, and roles, and the interplay among them has implications for the management of fisheries resources. It is possible that the alignment of these multiple actors will create diverse sources of social capital that can be useful for mediating the various problems that may face those involved in natural resource management. However, it is also possible that the existence of multiple actors and institutions will result in incompatible and contentious relationships that may have negative impacts on the sustainable management of fisheries resources.

At the community level, various user-actors may have contrasting objectives and interests, and their interactions may constrain participatory resource management. At Lake Jipe, actors that use fisheries resources at the community level can be categorised into resident and non-resident fishers. While resident fishers use fisheries resources in the area mainly for subsistence, non-residents carry out commercial fishing. These variations in objectives and interests result in conflictual relationships, especially when those actors who have a strong or permanent dependence on fisheries resources in the area perceive that the participation of other actors results in the degradation of the resources. These relationships may determine how time is used by resource managers. That is, instead of time being used primarily in enforcing sustainable resource use practises, it might be used to resolve the conflictual relationships among the users.

While governmental actors are entrusted with the facilitation of fisheries management, some of them may constrain the enforcement process. Fisheries management at the governmental level in Mwanga district is the shared jurisdiction of governmental actors at the district council level, as well as at the division, ward and village levels. Administrative and technical actors at the district, ward, and village levels have roles to play in the enforcement of participatory fisheries management. Although all these levels fall under the government, the existence and involvement of multiple governmental levels and actors is likely to complicate the tasks of enforcing, monitoring and coordinating fisheries policy, especially because actors may vary in their interests. The governance of the different jurisdictions may be rendered ineffective by tensions between self-interests and entrusted interests. While governmental actors at the higher levels (e.g. the district) may formulate regulations that have to be implemented at the local level, the actual implementation of these regulations depends on, inter alia, the commitment and willingness of actors at the lower governmental levels to enforce them. This will be explored further in the latter sections of this thesis, which analyse the participation of various governmental actors in the enforcement of fisheries management at Lake Jipe.

Having briefly highlighted the various actors involved in fisheries management at Lake Jipe, I will now analyse interactions between the government and the local community. This analysis is confined to the enforcement of fisheries management, conflicts and conflict resolution. Since, however, the enforcement of fisheries management is based on the implementation of the Fisheries Policy of 1997 and the Fisheries Act 22 of 2003, a brief overview of the fisheries policy and the act will be provided before the analysis of the interactions is presented.

4.2.5 Fisheries Policy of 1997 and Act 22 of 2003

Enforcement of sustainable fisheries management should be in congruent with the national Fisheries Policy of 1997, and the Fisheries Act 22 of 2003. These institutional instruments provide for community participation in fisheries management.

The establishment of the Fisheries Policy of 1997 necessitated a repeal of the Fisheries Act of 1970, and the revision of other fisheries regulations, namely the Territorial Sea and Exclusive Economic Zone Act of 1989, the Tanzania Fisheries Research Institute Act of 1980, and the Marine Parks and Reserve Act of 1994. This was because these acts were legacies of the former government-centred top-down management regime. The establishment of the Fisheries Policy of 1997 thus led to the enactment of the Fisheries Act 22 of 2003, which mandates participatory management of fisheries resources. Fishing communities in the fisheries resources areas are integrated in management plans and the implementation process.

The national Fisheries Policy of 1997 has many objectives, but the following are relevant to participatory fisheries management:

- Efficient use of the fisheries resource base to increase production and availability and to improve economic growth.
- Integration of conservation and sustainable management of fisheries resources in community social and economic plans.
- To promote the fishing community's involvement in the planning, development and management of fisheries resources.
- To strengthen cross-sectoral collaboration between fisheries and other sectors.
- To strengthen collaboration between fisheries and other sectors in the general development of the fisheries sector, and minimisation of operational conflicts.

These five objectives are discussed in greater detail below.

Efficient use of fisheries resource base

Collaboration among various actors is important to the efficient use of fisheries resources. The Fisheries Policy of 1997 highlights that for fisheries resource base to be used efficiently; there must be collaboration among various actors, including the user community, in the surveillance and enforcement of fisheries regulations. This is important for effective enforcement of the Fisheries Act 22 of 2003.

The Fisheries Act authorises the Minister of Fisheries to facilitate the enforcement of joint surveillance. The act underscores that joint surveillance involving the government, other relevant agencies, and the local fishing community is imperative for ensuring the enforcement of the act. Therefore, participatory management of fisheries resources between the government and the local people in the vicinity of the fisheries resources should be implemented. One of the important steps in creating joint surveillance is the formation of community management units. The community management units are seen as imperative to the sustainable development of fisheries resources. The formation of the community management units is emphasised in the act as a prerequisite for the appropriate management of fisheries resources. The act states that the use of community management units will guarantee the protection and conservation of fisheries resources because the primary stakeholders are integrated into the management regimes.

Although the act recognises the importance of the formation of community management units, it also places a high level of trust in the management of these units, but good management is not certain. The act does not foresee potential negative interactions between the management units and the community at large, and therefore, doesn't provide recommendations to mediate such problems. At Lake Jipe, for example, although the management units have been formed, they receive little cooperation from the community, who regard the units as government interference. This negatively affects participatory fisheries resource management.

Integration of conservation objectives with socioeconomic objectives

The policy understands that, in order to achieve sustainable fisheries management, an integration of social and economic objectives is needed in the conservation objectives. One of ways mentioned therein is the introduction of alternative options for generating livelihoods, such as establishment of fish farming (aquaculture). It highlights that through such considerations, it is possible to ban and control unsustainable exploitation of fisheries resources. At the Lake Jipe area, one of the reasons for the high dependence of fishers on fisheries resources is the lack of reliable alternative livelihood options. Alternative livelihood options could help in relieving high pressure on fisheries resources though the option mentioned in the policy (i.e. aquaculture) can be difficult to implement at Lake Jipe because this would require a reliable supply of water, which is a limitation at the Lake Jipe area.

Involvement of fishing community in fisheries management

The Fisheries Policy of 1997 views community participation in the management of fisheries resources as an important means for sustainable development. It sees fishing community associations (committees) as a pre-requisite for creating a sense of ownership in the management regimes, and therefore emphasises the formation of these associations. It urges various fisheries projects to involve the community in their implementation, as such involvement enables the empowerment of the fishing community. This policy also carefully identifies areas where the community will have incentives to participate. The policy encourages community involvement not only in the formulation of the policy, but also in its implementation. It does so by entrusting some management functions, such as monitoring landing sites and enforcing compliance with regulations requiring the use of sustainable fishing equipment, to the community.

The Fisheries Act also provides for community participation in fisheries management. In Article 17, Parts P, R, and S, the act emphasises the fundamental importance of the community's role. The act also gives the Fisheries Minister powers to impose conditions that are imperative for appropriate use of fisheries resources. These conditions entail the use of traditional practises, joint surveillance, and the formation of community management units.

The act views participation as contractually based. Part V, Article 18 addresses the agreement made between the fishing community and the government necessary to the participation of the local fishing communities in the management of fisheries resources. It highlights that the government, through the Director of Fisheries, is authorised to make such an agreement with the fishing community (beach management units). In this agreement, there must be a clear statement of objectives, a description of the agreed-upon area, a list of the management functions to be undertaken under the agreement, guarantees for the access and use of resources under the agreement by other users, a time period for the agreement, a provision for the revision of the agreement, and a provision for the settlement of disagreement. This indicates that the management of fisheries resources is considered the shared jurisdiction of the government and the community. Rather than applying top-down management, the government must negotiate with natural resources users for both sides to benefit from the management initiative. The act also indicates awareness that parties will disagree, and therefore methods for settling disagreements must be discussed.

The Fisheries Act recognises the traditional practises of the fishing community. The act authorises the responsible minister to ensure that traditional fisheries practises (which are consistent with sustainable fisheries management, needs and interests of local users' community highly depending on these resources) have to be respected and integrated into the management regimes. This ensures that there both scientific and local knowledge are integrated in the management of fisheries resources.

Strengthening of inter- and cross-sectoral collaboration

The Fisheries Policy recognises the importance of cross-sectoral collaboration in fisheries development. It states that collaboration among different sectors on various issues is important to addressing and minimising conflicts among different sectors. Consequently, it is also important to sustainable development of the fisheries sector. The policy suggests four methods through which the collaborations can be operationalised. These methods include programmes for joint meetings by all sectors related to environmental management, natural resources, tourism, and conservation of natural resources. All related sectors can use joint meetings to address overlapping issues that require multidisciplinary collaboration.

4.3 Enforcement of sustainable fisheries management practises

Enforcement of fisheries institutional instruments is important for sustainable management of fisheries resources and other resources that they depend upon. Enforcement ensures that, along with the socioeconomic concerns, environmental and sustainability aspects are considered. In this context, the natural resources from which socioeconomic benefits are derived are sustained be available today and in future. This section analyses the collaboration between the district government and the local community in the enforcement of sustainable fisheries practises.

4.3.1 Creating awareness on fisheries policy and legal instruments

An understanding of the institutional arrangements and boundaries of operation is crucial for effective enforcement of sustainable management practises. At Mwanga district, the district natural resources office creates stakeholder awareness concerning the Fisheries Policy of 1997 and the Fisheries Act 22 of 2003 (henceforth collectively referred to as fisheries policy). This is done through preparation of a seminar presenting the by-laws for enforcing sustainable fisheries management practises based on national fisheries policy. The seminar is given to the local key stakeholders to provide them with an understanding of areas of regulatory enforcement. Various relevant institutions are represented at the seminars, including the village community. This ameliorates barriers to implementation.

It is important that the whole community is aware of the legal instruments of resource management. In order to spread this awareness, representatives of the community participate in the seminars, and are then responsible for educating the rest of the community. At the village levels, such education is accomplished through the village assembly. There, local community members are informed about their responsibilities pertaining to fisheries resource management. Through this process, fisheries committees have been appointed in each village to steer the community in the management of fisheries resources.

4.3.2 Monitoring, control and surveillance

The government and the fishing committees share the responsibilities of monitoring, controlling and surveilling fisheries resources. However, in contrast to the former top-down approach whereby the government managed all of these functions alone (though its capacity to do so was limited), in the current arrangement, the local people assume the primary responsibility for day-to-day enforcement through the fisheries committees. The district government assists in these arrangements, especially through patrolling when fishers oppose the activities of the fisheries committees; for example, when fishers block the functions of the committees and threaten committee members. The district government also collects taxes, issues licences for fishing and fishing vessels, and ensures order in the monitoring, control and surveillance functions at the community level. Sometimes, surprise patrols are conducted when the district authority learns that unsustainable fishing practises are occurring at the lake area. This is facilitated through informal communications from agents in the villages on the presence of fishers involved in illegal fishing practises, and especially when the violators seem to overpower fishing committees in these villages. The police may become involved if necessary. In particular, police involvement takes place when fishers threaten the lives of the fisheries committee or hinder them from administering the jurisdiction entrusted to them.

Motivational and coercive mechanisms are used to confiscate unsustainable fishing equipment. The motivational approach offers fishers who own prohibited equipment a grace period for surrendering that equipment to fisheries authorities. If they do not comply, a patrol is carried out to arrest those who would did not surrender their equipment. Forceful confiscation, depending on the case, may be accompanied by legal charges. Table 4.2 shows the incidences of collaborative confiscation of illicit fishing tools by fisheries committees, and fisheries officers, village leaders, district fisheries officers and police in February and March of 2006 in the Mwanga district.

It shows that most of the confiscated tools were from Lake Jipe (6,302), and Mikocheni (2,000). The time period indicted is the time after which migratory fishers from Nyumba ya Mungu dam moved to Lake Jipe.

Consultative mechanisms are applied in addition to confiscation. The Natural Resources Department collaborates in District Commissioner meetings in various fishing communities and consults with them on how to strategically combat unsustainable fishing practises. These meetings are a response to unsustainable fishing practises and threats faced by the fisheries committee members when they enforce regulations. In these meetings, the District Commissioner discusses the government's stance against the violators of law and compromisers of peace. The District Natural Resources Department uses these moments to strategise with the community on how to contain unsustainable fishing equipment and practises.

Two main approaches to curbing unsustainable fishing practises are generally raised at these meetings. These are the encouragement of violators to comply willingly with regulations, and the use of coercive means. The encouragement of owners of prohibited fishing equipment to surrender that equipment is performed by designating a grace period for surrendering that equipment before applying the coercive approach (e.g. conducting a patrol from house to house, and/or involving the police in the confiscation exercise). From December 2005 to May 2006, confiscated equipment valued at a total of 24 million Tanzanian shillings was collected, as illustrated in Table 4.3.

In most cases compliance is limited, as indicated in Table 4.4 This necessitates the employment of forceful means through patrols that enforce compliance. For example, in 2006 (June to September) 56 articles of prohibited equipment were confiscated, of which, fifty (50) were small prohibited nets, while six (6) were mosquito nets.

Along with the confiscation of illicit fishing equipment, immature fish are also confiscated, and the violators have to pay fines for capturing juvenile fish. The fine is generally TZS 1000 (€0.63) per kg of small fish (Table 4.4). The monetary values indicated in Table 4.4 are the values of the confiscated tools and the fish as estimated by the fisheries officers at the district. The table indicates cooperation between the fisheries committee, the district natural resources department and the fisheries officers in patrolling to combat unsustainable fishing equipment. The participation of the natural resources department in the patrol occurs especially when the fisheries committee cannot implement these roles alone due to resistance from fishers, who sometimes even threaten their lives.

Once confiscated, the illicit fishing tools are burned. It is questionable however, as to whether burning the confiscated fish equipment is by itself a solution to curb illicit fishing. Although the prohibited fish equipment is confiscated and destroyed, such equipment reappears after a short period. This is not to say, however, that the practises of confiscating and burning the prohibited equipment are useless. The reappearance of the prohibited equipment nonetheless implies that such practises should be complemented with other practises that influence people's attitudes

Table 4.2. The implementation of a patrol to combat illegal fishing in Mwanga district for February and March in 2006 (source: Natural Resources and Fisheries Department, Mwanga district, 2007).

Date	Place/village	Prohibited gears confiscated	Actors participated in the patrol	Action taken against victims
10.2.06	Bora	2	• Lang'ata ward fisheries officer • Fishers committee	• Arrested and case opened at Mwanga police station
10.2.06	Njia Panda	3	• Fisheries officers at Kirya ward • Ward leadership	• Seized equipment was burned
17.2.06	Lake Jipe	2,625	• District fisheries officers • District police • Ward leaders • Fisheries committee	• Patrol of households • Seized equipment was burned
19.2.06	Njia Panda	18	• Kirya ward fisheries officer • Village leaders	• Seized equipment was burned • Owners escaped
19.2.06	Kiti/ Mungu	6	• Kirya ward fisheries officer • Village leaders	• Seized equipment was burned • Owners escaped
22.2.06	Bora	2	• Ward fisheries officer • Village fisheries committee	• Owners surrendered illegal equipment • Equipment burned
23.2.06	Kagongo	23	• Ward fisheries officer • Village fisheries committee	• Owners surrendered prohibited equipment • Equipment burned
24.2.06	Njia Panda	5	• Ward fisheries officer • Ward and village leaders	• Equipment burned • Owners hid themselves
11.3.06	Kagongo	3	• Ward fisheries officer • Fisheries committee	• 6 owners seized and case opened against them at Mwanga police station
11.3.06	Njia Panda	2	• Fisheries officers • Ward/village leaders	• Seized equipment was burned
15.3.06	Mikocheni	2,000	• District fisheries officers • Ward fisheries officers • Fisheries committee	• Seized equipment was burned
22.3.06	Kagongo	2	• Ward fisheries officer • Ward leaders	• 2 owners arrested and taken to police
23.3.06	Lake Jipe	3,677	• District fisheries officers • Lake Jipe ward leaders	• Patrol of households • Seized equipment burned • 1 owner caught, charges opened against him
Total		**7,955**		

Table 4.3. Confiscation of the destructive fish equipment at Mwanga district for December 2005 to May 2006.

	Prohibited fishnets	Mosquito nets	Total value (TZS)
Prohibited fishing equipment returned willingly	63		
Prohibited fishing equipment returned during the patrol	2,078	11	
Total	**2,141**	**11**	**24,050,000**

Table 4.4. Confiscated illegal fish and fishing equipment from June to September 2006 in the Mwanga district (source: Mwanga natural resources office, 2007).

Date	Illegal practise uncovered	Monetary value (TZS)	Participated in patrol
June 2006	23 prohibited fishnets seized	14,320,000	Ward fisheries officer Fishers and fisheries committee
July 2006	73.5 kg of juvenile fish confiscated	420,000	Natural resources officer
	22 prohibited fishing nets and 1 mosquito net seized	13,080,000	Ward fisheries officer Fisheries and fishers committee
Aug-06	17 kg of Juvenile fish seized	15,000	Fisheries committee
	1 prohibited fishing net and 5 mosquito nets	1,050,000	Ward fisheries officers
Sep-06	35 kg of juvenile fish seized	35,000	District natural resource officers
	4 prohibited fishing net seized	2,600,000	Fisheries officers and fishers
Total		**31,520,000**	

towards resources. At Lake Jipe, for example, fishers remarked that they were never trained on sustainable fishing practises, but have been chastised for conducting illegal fishing practises. The fishers' claims were confirmed during my interview with the District Fisheries Officer. He remarked that fishers know the effects of their activities and would act in a similar way even if they were trained. This implies that no capacity-building seminars have been given to the fishers. Again, the reappearance of illegal fishing equipment after destruction indicates that the problem has only been partially addressed. It appears that the focus is on the product, while the cause is not addressed. Illegal equipment may be sold by markets or fishers may modify legal equipment to allow them to catch smaller sizes of fish. The current co-management arrangement therefore also needs to consider the sources of destructive fishing equipment, so that the sources of the problem, as well as its effects, are controlled.

In natural resource management, when carrot approaches fail, stick approaches are used. The use of police has been already introduced as a method adopted by the fisheries department at Mwanga district to enforce compliance with fisheries legislation. The police are involved when the fisheries patrolling team (which may involve fisheries officers alone or fisheries officers and committee members) encounter life-threatening resistance from the fishers, as indicated in the following report from the Ward Fisheries Officer:

> *When we were conducting a patrol at around 9:30 AM, I, together with the Ward Executive Officer (WEO), Village Executive Officer (VEO), Auxiliary Police, and a group of 20 fishers (fisheries committee) seized three persons who had fished juvenile fish along the lake shore. We arrested them. One of them made a traditional emergency shout for help, traditionally called ukunga, and, to our surprise, a group of people came with stones, swords and big wooden cooking spoons in their hands. They started stoning us, and one of the violators bit a person who was holding him and managed to escape. We could not continue with the patrol, but managed to leave with two violators, and we took with us the evidence of juvenile fish to the nearby police post. While there, we called the District Natural Resources Officer and he joined us at the police station. The Natural Resources Officer requested to be provided with three armed police officers to accompany us to the village, where we continued the patrol and arrested other violators with prohibited fishnets and mosquito nets and led them all together to the central district police station to open a legal case against them (Ward Fisheries Officer, Kirya ward, July 2006).*

The quote above shows how difficult it is for the fisheries committee and the fisheries officers to enforce sustainable fisheries practises in some villages. It indicates that the fishers in these villages collaborate against the enforcement of sustainable management practises. They even use traditional mechanisms and networks such as *ukunga* (danger alarm) to fight against the formal institutional arrangements. This necessitates the use of force because the safety of those responsible for enforcing sustainable fisheries regulations enforcement is at stake. In these villages, the appointed fisheries institutions have less power to counteract unsustainable fishing practises and, patrolling therefore has to involve the police. The use of the police is in keeping with Fisheries Act 22 of 2003 as provided in Part 4, Section 37. In such a situation, however, the sustainability of management initiatives is questionable, since it is costly to carry out this practise regularly, and the local people may not assume ownership of resource management.

Some responsibilities are entrusted to the government. These include pressing charges for non-compliance with fisheries regulations, raising of awareness among the users of the fisheries resources, and inspection of fish and fining for fishing and transporting juvenile fish.

The Natural Resources Department presses legal charges against violators of fisheries regulations. Legal charges are especially common when an individual or group of fishers intentionally hurts those responsible for enforcing sustainable fisheries management practises. For example, in 2006, the fishers beat and wounded a fisheries officer at Nyumba ya Mungu dam when he was confiscating prohibited fishing equipment. They dislocated his arm and wounded his liver. Charges are also common when fishers fail to pay the fines on equipment that has been seized. When legal charges are pressed against violators, they are punished in accordance with the Fisheries Act 22 of 2003. If a person is convicted under this violation, the act provides for a prison sentence of not more than 6 years (Part IX, Section 41 of the act).

Creating awareness of the legal instruments of fisheries sustainability is important to the active participation of relevant actors. At Mwanga, when the participatory fisheries management approach was established, the District Natural Resources Office raised stakeholders' awareness on national fisheries policy through seminars. This was imperative to ameliorating barriers to implementation. Various relevant institutions were represented, including the fisheries committees and government leaders at the local level. The representatives, thereafter, were responsible for creating awareness of fisheries policy among the rest of the community and other relevant actors. At the village levels, such awareness was provided through the village assembly. The local communities in such meetings were informed about their responsibilities to fisheries resource management. It is through this process that fisheries committees were appointed in each village to steer the community in the management of fisheries resources.

The inspection of fish transported for trade in urban areas from the villages is designed to control fishing of juvenile fish. According to the Fisheries Act 22 of 2003, fishing of small fish is prohibited. The Mwanga district fishing by-laws prohibit fishing of fish of less than two inches. When caught with immature fish, violators are fined, as indicated in Table 4.5. On some occasions, this practise has led to a clash between the natural resources auxiliary policies and the fishers at the checkpoints. For example, one time when the researcher was travelling from Jipe village to Mwanga town, at the inspection point, the fish businesspersons and the natural resources auxiliary police fought because the fish businesspersons resisted paying the fine when found guilty after inspection. In order to minimise risks to them, fish businesspersons sometimes avoid checkpoints by transporting fish at night, when the checkpoints are not in operation.

Table 4.5 indicates that for every kg of small fish confiscated, the victim is fined TZS 1000. The Mwanga District Council has modified the penalty and amounts stipulated in the Fisheries Act 22 of 2003, Article 47. According to the by-laws, 100,000 TZS is charged as a fine for catching

Table 4.5. Fines charged due to fishing and transportation of small sized fish at Mwanga district (1000 TZS = 0.63 €).

Date	Violator	Case/violation	Fine (TZS)
11/7/07	Fisher 1	Transportation of small fish for sale less than 2 inches (30 kg)	30,000
22/6/07	Fisher 2	Transportation of small fish under 2 inches (100 kg)	100,000
18/9/06	Fisher 3	Fish under 2 inches (10 kg)	10,000
18/9/06	Fisher 4	Fish under 2 inches (10 kg)	10,000
18/9/06	Fisher 5	Fish under 2 inches (10 kg)	10,000
18/9/06	Fisher 6	Fish under 2 inches (5 kg)	5,000

fish smaller than the two inch minimum, but it did not mention any specific scale. The Natural Resources Office at the district has modified this by charging according to the weight of small fish that are seized.

This section discussed the participation of the government and the local community in the enforcement of sustainable fisheries practises. It described the main activities undertaken by those responsible for enforcing sustainable policy, including creating awareness of fisheries policy. It also described other practises undertaken by the government and community actors in the enforcement process, which are monitoring, control and surveillance, confiscation and destruction of illicit fishing equipment, confiscation of small fish, and fining. It indicated that in undertaking the monitoring, control and surveillance functions, the government and the local community, as represented by the fisheries committee, collaborate. Although the central day-to-day roles are undertaken by the fisheries committees, there are instances when the fishers resist the fisheries committees. The committees then require help from the government through the ward fisheries and the district fisheries officers. Further, in some instances, the government fisheries authorities and the fisheries committees cannot perform their enforcement functions alone because the fishers threaten their lives. At this point, they need help from the security police because the auxiliary police at the village levels fail to contain the resistance of the fishers. Confiscation of fishing equipment is also difficult for the fisheries committees to perform alone. Although this has been possible in some villages, in others it has resulted in direct harm to those enforcing the regulations. In these cases, the involvement of the District Natural Resources Department and the security police was necessary to confiscate unsustainable fishing tools. In addition, this section has indicated that, along with confiscation of illicit fishing equipment, small fish are also confiscated, and violators are charged fines for catching or transporting small fish. This affects both fishers and fish businesspersons transporting the fish to markets, who have to pay a fine if they are caught carrying small fish. However, as discussed above, this sometimes involves clashes between the natural resources auxiliary police at the checkpoints and the businesspersons.

The foregoing discussion has indicated that, in general, there is good cooperation between the government and the fisheries committee at Lake Jipe in combating unsustainable fisheries management practises. This is not always the case, however. Sometimes leaders at certain levels have not been sufficiently cooperative or have been unable to facilitate the activities of the committee. This has resulted in conflicts between the fisheries committee and governmental leaders. At other times, constraints at the governmental led to a failure to meet some responsibilities, especially at the district level. When the fishers experienced such a hardship, they resolved to employ specific strategies in attempt to address the situation. The aim of the coming section is to analyse these strategies the fisheries committee undertakes.

4.3.3 Strategies to counteract constraints of sustainable fisheries management

This section analyses different strategies of the fisheries committee when it is blocked by the fishers and the fishing community, or when the enforcing actors at the governmental level do not cooperate with the fisheries committees or do not facilitate their work as is required under the institutional and legal framework.

Fisheries committees respond to the poor performance of facilitators at local level by bypassing the next reporting level in the chain of authority. For example, if fisheries officers and administrators at the village and ward levels do not facilitate the work of the fisheries committee, the committee bypasses them in the reporting chain. The committee itself presents the problem to the district fisheries office. In other words, bureaucratic protocols of accountability are ignored in such situations. A section of a letter from the District Fisheries Officer (DFO) to the Ward Executive Officer (WEO) illustrates this response:

> *We have received complaints from the Secretary of Fisheries committee at Lake Jipe regarding unsustainable fishing practises. On 4/6/06, the committee patrolled around the lake, and illicit fishers blocked them. They reported to your office but you did not give them assistance. You may recall that we previously patrolled around Lake Jipe and confiscated and burned illicit fishing equipment. Therefore, we expected that your office and fisheries committee would not allow the illicit practises to re-emerge. Due to the deficit of fisheries officers, we have only one officer in the division stationed at Kigonigoni village. For the successful protection of fisheries resources against illicit fishing, the cooperation of your office and the fisheries committee is necessary (District Fisheries Officer, for the District Natural Resource Officer, 10/6/2006).*

The letter shows how hierarchical government structures can be bypassed by the fisheries committee when the hierarchy acts as an obstacle to enforcing sustainable fisheries management. The letter indicates that the District Natural Resources Office received complaints from the fisheries committee on the laxity of the leaders at the village and ward levels in curbing illicit practises. The district authority at the district intervened by ordering village and ward actors to cooperate with the committee.

Direct communication with the District Commissioner is another method used by the fisheries committees to empower their organisations against unsustainable fishing practises and against seemingly non-cooperative village and ward leaders. At the Jipe-Ndea division, fisheries committees have used this method to form a divisional fisheries union. This self-organisation is aimed at strengthening their voice in their concerns against illicit fishing and the laxity of enforcement instruments from the government at the village, ward and division levels. This action resulted from the realisation that technical officers entrusted them to perform functions that were not facilitated by the government. Furthermore, this organisation is designed to collaborate against illicit fishers who do not comply with fisheries regulations; when they violate regulations in one village, they flee into another. That is, after they have clashed with the committee and their equipment has been confiscated, fishers move to another village where they proceed with their unsustainable practises. Since the fisheries committees in the villages had been working independently of other similar committees in other villages, they could not address this situation adequately. The divisional fisheries committee union is designed to counteract this problem. This is illustrated in the letter from the union addressed to the District Commissioner, accusing the technical and administrative staff of not cooperating with the committee. At the same time, the letter informs the District Commissioner that they have formed a union at the division level, and they ask him to approve it. A portion of this letter is included below:

We would like to inform your office about illicit fishing at Lake Jipe. We the fisheries committee have decided once again to inform your office about this situation, which is worsening daily. Last year, we informed your office about this situation and your office instructed the divisional secretary of the Jipe-Ndea division to take charge of this issue of illegal fishing. But what surprises us is that since the divisional secretary was given your instructions, he has not implemented even a single directive and now the situation is worsening. The situation at Lake Jipe is becoming extremely poor because since last year, many fishers from Nyumba ya Mungu dam in Mwanga and Manga dam in Korogwe have come to Lake Jipe, and these fishers are very experienced at the illicit capture of juvenile fish using dragnets and small netted fishnets of 1.25". Honourable DC, we support your sincere efforts, which you have started implementing at Nyumba ya Mungu dam to combat illegal fishing, and we ask you to introduce such initiatives to Lake Jipe. Hon. DC, this committee would like to inform you of one thing. Many past DCs of this district failed to achieve their goals of combating illegal fishing because the village, ward and divisional leaders would not give them cooperation and this has even led to the loss of possible revenues. We need your assistance because here at Lake Jipe we do not have a fisheries officer. We have heard that the district sent a fisheries officer to this area last year, but we have never seen him. We ask you to take this issue seriously. Last year, we came to your office as a village fisheries committee for Jipe village, but now we have formed a divisional fisheries committee because Lake Jipe is found in two wards, Jipe and Kwakoa that together compose the Jipe-Ndea division. We have formed this divisional committee without informing the divisional secretary because he has never cooperated with the fisheries committees in the two wards. We need to be empowered to arrest illegal fishers and take them to legal facilities (Jipe-Ndea Division, fisheries committees, 2007).

Procedurally, when such union is formed, leaders and technicians, and the village, ward and district levels must be informed. Since the fisheries committees have identified the leaders who seem to be obstacles to their roles, however, they bypassed them and pursued action from the higher authority at the district level.

4.3.4 Community participation in fisheries management

Although the fisheries committee represents the community, community members still have specific responsibilities in the management of fisheries resources. The role of the community can contribute to the success or failure of fisheries management as a whole. The participation of the community in the management of fisheries resources at Jipe mainly involves appointing fisheries committees. Although the community appoints the committees, it does not provide them with due cooperation at the policy enforcement stage. The committees are regarded as policing institutions that stand for the district and national hierarchies, but not as an integral part of the fishing community. Though the committees are elected by the fishers, they are regarded as impositions from the central and district governments. Such thinking is perhaps premised on the fact that the functions performed by these committees are similar to those performed by the district officials before the formation of these committees – surveillance, monitoring, and control. Even in the current management regime, the District Natural Resources Department facilitates the committees, especially when blocked to implement their jurisdictions.

The fishing community regards the ownership of fisheries committees as existing outside the boundaries of the fishing communities. There is evidence that before the committees were appointed, insufficient effort was exerted to make local people clearly understand the rationale of forming these committees and whom they were going to serve. It also seems that less knowledge exists on the governing institutions in some parts of Lake Jipe community. The fishing community may compel the committee to work in the way they deem proper, and failure to do so would result in voting members off the committee. This is the case especially when catches from the lake decline and some fishers tell the community that the reason for this problem is the strictness of some members of the fisheries committee. This generally indicates that fisheries policy and regulations are not well understood among the fishing community.

Although the fisheries committee has the power of enforcing fisheries by-laws, the community has the power of deciding who should be included in the committee. Some elites know this and intelligently use the community to remove committee members who seem to be barriers to their interests. This is illustrated in an interview with a member of the fisheries committee in Butu village (a former chairperson). He was dismissed from his chair position because he was strict at enforcing by-laws against illicit fishing practises, while the local people wanted him to slacken the by-laws. When he was persistent, they replaced him with another chair.

> *The by-laws explained plainly that fishing nets of less than two inches should not be used for fishing. When I was chairperson of the fisheries committee, I had to make sure that this regulation was followed. Later, I discovered that when you become strict, people hate you. However, I think the fishers who carry out illicit fishing practises mobilised the people to force my dismissal from the chair position. People started to force me to allow them to use small fishnets, which I did not agree with. To my surprise, one day they organised a meeting and told me that from that day on I was not the chairperson but a mere member in the fishers committee (Member, Fisheries Committee, Butu village, 2007).*

The current management framework empowers the community to determine the fate of members of fisheries committees because they appoint them. A lack of mechanisms to safeguard the security of committee members is an obstacle to the committee's smooth functioning. There is currently no formal document that codifies the mode of membership in the fisheries committees and conditions for termination. This enables the community to remove members from the committee without even a rational analysis of the situation. Members may thus be displaced even when they operate within a framework of stipulated regulations.

The foregoing discussion describes how actors have different interests, which in one way or another trigger them to act at an expense of other actors, and also against sustainable resource management concerns. This demonstrates that in natural resource management, where many actors with different interests and intentions are involved, conflicts are inevitable.

4.4 Conflicts and conflict resolution in fisheries management

This section analyses conflicts that emerge in the management of fisheries resources at Mwanga district in general, and Lake Jipe in particular. It also discusses approaches used to mediate and resolve these conflicts.

4.4.1 Conflicts

The following first analyses conflicts at the community level that emerge in the management and use of fisheries resources among various actors. In this respects, conflicts at this scale are termed *intra-community conflicts*. These conflicts occur at the village, ward and division governmental levels. Conflicts between the district government and the community, termed *cross-scale conflicts*, are then addressed. Finally, conflicts that arise across national borders, between fishers on the Tanzanian side and Tsavo National Park authorities on the Kenyan side, are discussed. This type of conflict are termed *cross-border conflicts*. Table 4.6 summarises these conflicts.

Intra-community conflicts

As introduced above, conflicts within the village community occur between the village/ward/ division government and the fisheries committee, between resident and non-resident fishers, between the fisheries committee and fishers, between village governments and non-resident fishers.

In collaborative decision-making, the failure of one side to fulfil its entrusted responsibilities constrains the effectiveness of another side, and may result in conflicts between collaborating parts. At Lake Jipe, the village, ward and division governmental leaders all engage in conflicts with the fisheries committee when these government leaders become an obstacle to the enforcement of fisheries by-laws. Based on participatory fisheries resource management, the committee has to collaborate with the village, ward and division governmental levels. Governmental actors are responsible for facilitating the committee by, for example, inflicting punitive measures upon fishers who threaten the committee and block them from implementing their responsibilities. Some government leaders fulfil their responsibilities. The most egregious violation of the governments' role is when leaders ally with those involved in illicit practises. The fisheries committee is aware of when leaders are likely to benefit from illicit income obtained from violators, and responds with distrust. This lack of trust creates a lack of cooperation. The fisheries committee is then inclined to bypass the governing structures at the village, ward and division levels, and advances its concerns to the district governmental level. Thus, the committee demonstrates its belief that local governmental actors cannot facilitate its work. It therefore looks to governmental actors outside the village boundary for support. This weakens collaborative natural resource management. The successful management of natural resources requires cooperation among actors at different positions.

Irresponsibility contributes to the constraints on sustainable fisheries management. In this context, by irresponsibility I mean that those expected to facilitate community empowerment, capacity building and other facilitatory responsibilities, do not fulfil those responsibilities. Two aspects are relevant in the irresponsibility domain: namely, administrative and technical.

Table 4.6. Conflicts among fisheries actors as one of the problems of the fisheries sector at the Lake Jipe area (source: field survey, 2006).

Actors involved	Source of conflict	Resolving institution (s)
Resident fishers versus non-resident fishers	• Use of destructive equipment by non-resident fishers • Distortion and removal of resident fishers' equipment by non-resident fishers	• Village government • Natural Resource Office • Elders • Divisional Secretary
Fisheries committee versus fishers	• Illicit fishing practises • Not observing the established institutional arrangements	• District Natural Resources Department • Police • Ward Fisheries Officers • Ward Executive Officer
Non-resident fishers versus village leaders	• Use of illegal fishing equipment • Non-resident fishers threaten to physically assault the leaders • Non-residents fishers do not observe village rules	• Village government and established by-laws • General village meetings
District/Division/Ward/Village leaders versus members of the fisheries committee	• Not recognising the role of the committee, interfering in negative way • Not facilitating and protecting the committee against illicit fishers • Controversially releasing the violators of regulations	• District Commissioner • District Natural Resources Office
Tsavo national park officers and Kenyan police versus Tanzanian fishers	• Tanzanian fishers invading the Tsavo National Park	• Legal actions taken, dialogue between Tz and Kenyan governments

Administratively, it is the responsibility of administrators at the local level (e.g. ward executive officer, divisional secretary, village executive officers, etc.) to intervene when order and peace is compromised among local actors in the policy implementation process. This is intended to safeguard that the assigned responsibilities are fulfilled. The lower community institutions for enforcing fisheries regulations require cooperation from government facilitators. The administrators have the job of creating a harmonious and orderly environment to avoid uncertainty among community actors when performing their entrusted roles. By the same token, the extension service providers are expected to ensure the provision of relevant services, advice and training that enhance the capacity of the enforcing committees at the community level. The failure of governmental actors to appropriately fulfil their responsibilities constrains the committees in their work.

In other situations, government leaders, who should facilitate and support the enforcement of fisheries management by-laws, may oppose the process. These leaders align with illicit fishers and obstruct the action of the fisheries committee and responsive leaders. The following portion of a letter from the Ward Fisheries Officer to the District Fisheries Officer illustrates this phenomenon.

> *Refer to the heading above. I am informing you that the illicit fishing that continues at Kiti cha Mungu is supported by village leaders. On 12/06/2007 at 5:30 pm, I, with the Ward Executive Officer (WEO) and some fishers, conducted a patrol at the area where illicit fishing equipment was used, and we found three boats with ropes for pulling illicit fishing nets. When we were burning the ropes, a group of fishers appeared and insulted us, claiming that we were harassing them. Some of them assaulted a member of the fisheries committee who was with us in the patrol, claiming that he is too strict, and that he had better die. This fisher also stated that no one among us could arrest him, but that only the police could do that. They continued to threaten us, and later one of them asked us 'why are you leaders very strict, not like our village leaders?' This statement is similar to claims we have heard from resident fishers who refuse to participate in patrols until their village leaders participate because the illicit fishers give the leaders money every month, so that they do not monitor them but instead protect them. In total, those who were insulting and threatening us were five fishers. They claimed that they would kill two leaders of the fisheries committee, and if they failed to do so, they would set their houses on fire. These threats have been reported to a local police station (Kirya Ward Fisheries Officer; Mwanga District 17/06/2007).*

The letter indicates that some leaders constrain collaborations between the government and community institutions (i.e. the fisheries committee) in enforcing sustainable fisheries management by promoting illicit gains and defending the law violators.

Competition for scarce resources is another source of conflict among resource users in the community. When a resource is scarce, each group strives to procure a part of that resource, and in that struggle, the interests of other users may be ignored or affected. One such conflict is that between non-resident fishers and resident fishers at Lake Jipe. There is a general conception among Lake Jipe residents that non-resident fishers, besides carrying out unsustainable fishing practises, are cruel. At Ruru subvillage, for example, non-resident fishers dragged their fishing nets in fishing sites where the residents had set their nets, hooks, and fishing baskets. This kind of conflict was intense because it led to the use of knives and spears. Had it not been for the elders' intervention, the conflict could have resulted in fatal outcomes. The elders and the village government then reported this conflict to the Natural Resources Department, and police at the district. The District Natural Resources Office intervened to resolve these conflicts by exhorting the fishers to respect one another's fishing arrangements.

Conflicts between resident and non-resident fishers are not only attributed to the interference of fishing sites of resident fishers by non-resident fishers. Generally, the two conflicting groups are not on good terms. It is broadly true that if resident fishers had the power to prevent non-residents from fishing in Lake Jipe, they would do so. Such a power, nonetheless, does not exist, so the two have to use fisheries resources together. This is because the resources are owned by the government, not by resident fishers.

The existing fisheries legal instruments do not put boundaries on resource users. According to these procedures, anybody can fish anywhere in the country (Tanzania) if he/she has procured fishing and fishing vessel licences, and pays the required taxes to relevant authorities. From a sustainability point of view, this makes it difficult to control unsustainable fishing practises. For example, it is difficult to get migratory users to engage in sustainable practises because their interest is to obtain adequate amounts of fish, but not to sustain the resources of the lake. This is because when fishing resources are depleted in one area, migratory fishers move to another. This results in conflicts between migratory users and the residents of the particular area, as the residents depend on the resource on a daily basis for food and income. For residents, Lake Jipe has been a major source of their livelihood for many years. They are therefore concerned about migratory fishers' degradation of that resource. This concern is so strong that non-residents are excluded from fisheries committees on the argument that their inclusion could lead to greater degradation of fisheries resources. The residents assume that the inclusion of non-resident fishers in the committee would give them the opportunity to perpetuate poor fishing practises.

However, non-resident fishers who have intermarried in the resident community and have established settlements are now regarded as a part of the resident community. The exclusion from the fisheries committees therefore does not apply to them. For example, there are fishers from Lake Nyasa in the southern regions of Tanzania and from the Singida region in central Tanzania who are members of the fisheries committees. These individuals are regarded to have the same stake in fisheries resources as the *Pare* residents because they have become a part of that group. By excluding non-resident fishers from the fisheries committee, resident fishers and the community imply that they have a greater commitment to the lake than non-resident fishers. In other words, they believe that they are to be more likely affected by degradation of the fisheries resources than non-resident fishers. This belief most likely originates in their persistent dependence on the lake for subsistence and income.

The scarcity of dependable resources creates pressure to relax the enforcement of governing rules, which leads to intra-community conflicts between law enforcers and resource users. At Lake Jipe, this conflict exists between the fishing community and fisheries committee. When fish supplies are scarce, the fishing communities in some villages want the fisheries committee to be less strict in its enforcement of fishing regulations. This was the case in Butu village, where a member of the fisheries committee was dismissed from the chair position because he was strict in enforcing the regulations when fish were scarce. Since the village community appoints the fisheries committee and, therefore, can dismiss them, they use this power to enforce adaptations of the regulations based on the fish availability. In other words, some fishing communities want the regulations to be implemented on the basis of seasonal fish availability. When there is abundant fish, the regulations can be enforced in keeping with established institutional arrangements, whereas when there is scarcity, they should be less strict.

This may imply that when fish are scarce, even fish of sizes smaller than recommended (the recommended size is at least 2 inches), could be harvested. The local people consider resources conservation to have lower priority when socioeconomic needs are not met, especially in times of deficit. They try to cope with this scarcity by enforcing the modification of the formally established arrangements to counteract seasonal shortages and meet their needs. When the limited livelihood options are less productive, the priority of environmental concerns becomes low. Balancing the conflicts of interests between socioeconomic and environmental concerns at such a time is a challenge.

Fisheries committees and fishers also conflict when the latter do not want to comply with orders from fisheries committees requiring them to stop illicit fishing practises. A portion of a letter written by the Ward Fisheries Officer to the District Fisheries Officer illustrates this fact:

> *Today, Sunday, a member of the fisheries committee prevented two fishers from fishing using small fishing nets and mosquito nets. They decided to assault him. While assaulting the committee member, a religious leader, who was passing by, asked them to stop the assault. One of the fishers, the owner of illicit fishing equipment, beat the religious leader using either a knife or a sharp edged tool and wounded him on his head, so he was bleeding. They also took off the committee member's shirt, beat him, and threatened that they would set fire to his grass-thatched house. The wounded reported to the police station. At the same time, the illicit fishers have hidden their boat and illicit fishing equipment in a place we do not know. These fishers have always been stubborn and they do not even want to surrender their illicit fishing equipment (Ward Fisheries Officer a Kirya Ward, 19/2/2006).*

The consequences of the conflicts have the potential to extend beyond actors directly engaged in enforcement. In the case above, a mediating actor suffered in the same way as the member of fisheries committee. Tension between socioeconomic and ecological interests caused this consequences. There is a strong tension between the procurement of livelihoods and environmental concerns. One reason for this tension is that the fishers do not have alternative livelihood options to diversify sources of food and income. They do not, therefore, consider the effects of their practises today against the availability of resources in the future. The reduction in the sizes of fish obtainable from the lake due to their own unsustainable harvest methods triggers further use of unsustainable fishing tools to obtain substantial amounts for subsistence and income. This constrains the efforts of the fisheries committee. Collaboration with the District Natural Resources Office and the security enforcers, i.e. the police, is thus sought to force the violators to comply.

Non-observance of established institutional arrangements leads to conflicts. This may be caused by simple ignorance, or by a strong desire to fulfil livelihood needs. The lack of diverse livelihood options may also catalyse this non-observance. At Jipe, conflicts occur between the village government and non-resident fishers. This conflict occurs because non-resident fishers do not recognise the governing and fishing institutions in the hosting villages. In addition, non-resident fishers threaten village leaders. The leaders report this in ward development committee (WDC) meetings, which are regularly held every quarter of the year. For example, in the WDC meeting held on 11th May 2005 at Jipe village, the village executive officer (VEO) reported that non-resident fishers threaten the leaders when required to comply with the governing rules. They do not want to follow the village rules and fisheries by-laws.

Cross scale conflicts

Cross-scale conflicts occur between fisheries committees and District Natural Resources Department. The fisheries committees, as previously indicated, are keen on performing the responsibilities entrusted to them. This justifies their efforts to counteract unsustainable fishing practises. They demonstrate their commitment to their responsibilities by their practise of bypassing some governing structures when those structures do not effectively act against illicit fishing practises. A portion of a letter written by the fisheries committee, and addressed to the District Commissioner illustrates this reality:

> *Refer to our letter Ref. S/R/No. A01/2006 we addressed to the Division Fisheries Officer of Jipe-Ndea Division on the disputable release of illicit fishers we had seized. After the release, these fishers are carrying out even more destructive fishing methods using illicit equipment. We ask your office to empower us to deal with this issue (Fishers committee, Ruru subvillage, 13/2/2007).*

The source of the conflict in the case above is the fisheries committee's surprise at the controversial release of the illicit fishers they had arrested. It seems that the fisheries committee was not satisfied with this release. What surprised them further is that after their release, the violators once again engaged in the same violation, only even more intensely. It is possible to imply from the situation that violators of the regulations might have bribed village and ward level governmental actors. The fishers, having seen this weakness, follow the established bureaucratic procedures by informing the next higher government hierarchy, that is, the Division Fisheries Officer. However, silence from the technical officer at the division level ended their pursuit of this matter. They then reported the issue to the District Commissioner (DC). It may seem surprising that the fisheries committee did not go through District Natural Resources Office, but their previous experiences whereby similar concerns were not appropriately addressed by that officer might have triggered their decision to report to the DC directly.

Prior to writing this letter, the fisheries committee had requested assistance to curb unsustainable fishing practises. The assistance was sought when the committee observed that their lives were threatened by the illicit fishers. They asked the District Natural Resources Office to bring the police in to assist them in confiscating prohibited equipment and small fish already harvested. The Natural Resources Office responded that it was constrained by financial resource. Surprisingly, the committee saw members of that office visit the villages to collect taxes from fishers. Thus, they interpreted this to mean that the Fisheries Officers were likely to be involved in illicit arrangements with the illicit fishers. They therefore decided to report this in writing to the DC. They wrote a letter accusing the Natural Resources Department and asking the DC for assistance. A portion of their letter to the DC affirms this:

> *We first of all commend you for your efforts to control illicit fishing. Honourable DC, illicit fishing has tremendously increased and now it is done openly at any time. Our lives are endangered as members of fisheries committee, as are the lives of all the people who support us. Owners of prohibited fishing equipment and fishers have*

established a squad to harm us. Honourable DC, we have reported this situation to the District Natural Resources Office so that they would request the Commanding District Police (OCD) to send policemen to our village to assist us with the exercise of confiscating and destroying prohibited equipment and arresting the violators. The Natural Resource Office has ignored our requests. We ask you to assist us with ten police officers for us to be secure in controlling the illicit fishing practises. Honourable DC, the District Natural Resources Officers are not with us in fighting against illegal fishing, they work for their own interests. When we ask them to bring the police, they respond that they do not have fuel for transport, but surprisingly they come daily to our village to issue fishing licenses. We cannot understand whether the problem is fuel or that the owners of the prohibited fishing equipment have bribed them. We ask you, Honourable DC, to assist in solving our dilemma. In addition, we request you to assist the Ward Fisheries Officer of Lang'ata by providing reliable transportation for him to be able to patrol the entire ward. Presently, it is hard for him to move around this vast area on foot. So, he is supported with a bicycle given to him by Good Samaritan villagers (Fisheries committee Kigongo village, 5/9/2006).

The account above illustrates the reason why the fisheries committee overlooked the District Natural Resources Office. This was based on their previous experience, whereby this governmental department could not help in addressing problems the committee faced in its work even as the committee was convinced that department had the power to do so.

It would be possible for the District Natural Resources Officers to misinterpret the committee's complaints as being attributable to the villagers' lack of understanding of the financial constraints experienced by the district department. Nevertheless, the District Natural Resources Officers need to acknowledge that the committee is doing a good job, and that their concern demonstrates that are committed to undertaking sustainable management of the fisheries resources even by assessing the accountability of the district authority. This attribute can be developed further so as to improve sustainable management of fisheries resources in the area. Care should be taken not to lower the morale of fisheries committees. In order to sustain the morale of the committee, relevant governmental authorities need to act accordingly when the fisheries committee presents issues that hinder the sustainable management of fisheries resources in the area.

The failure of the District Natural Resource Officer to appropriately execute his responsibilities is sometimes due to reasons beyond his control. Limited financial capacity is among the restrictions that influence the implementation of fisheries policy at Mwanga district. This problem renders institutions and departments entrusted with facilitation ineffective. Coupled with information gaps regarding the powers of higher departments between facilitators and the entrusted executers of specific policy roles, this results in conflicts and mistrust between involved actors in certain shared roles. This is the case in the interactions between Fisheries Department at the Mwanga district level and the fisheries committee, whereby the former is expected to facilitate the later in ensuring their security in surveillance and monitoring responsibilities.

Contradictory messages from leaders also cause cross-scale conflicts. Messages from the leaders to their constituents may cause an impasse in fisheries management. While authorised institutions enforce sustainable management practises and by-laws, other resource users may use the messages delivered to the community as rationale for their non-compliance. At Mwanga, this problem has occurred between the fishers of some communities and the ward and the district fisheries officers. The cause of this impasse was a speech, which was delivered by the Minister for Natural Resources who was interpreted as having said that fishers should fish without fishing licences. The following section of a letter from District Natural Resource Officer to the District Executive Director (DED) indicates this:

> *The patrol was planned because illicit fishing resumed in the Nyabinda and Njia Panda villages. The illegal fishing included the use of prohibited fishing equipment such as small sized dragnets, and fishing without a fishing license, which is a violation of fishing laws of 2003 and the regulation of 2005. We requested 10 police officers for our security, but were given seven. When we arrived at the landing sites at the lake, it was around the time for fishing, and fishers were about to pick their vessels and tools to get in the lake. We announced that we had to check that only those with fishing licenses could be allowed to pick their vessels and get in the lake. Suddenly, a big group of fishers with stones and clubs in their hands surrounded us. They blocked our escape route with big logs, and threatened to shed our blood. We tried to persuade them not to harm us and calm down but they would not. The fishers then claimed that they have been allowed to fish without licenses by the Minister of Natural Resources and Tourism, and demanded that we should bring him so that he could discuss this with them. We ultimately decided to leave the area because our lives were endangered (Natural Resources Officer, Mwanga District, 21/11/2007).*

In order to gain acceptance from constituents, and to win their votes for the upcoming elections, political leaders may deliver speeches to show they are responsible, caring, and sympathise with problems of their people, and want to assist them to achieve solutions to their priority problems. Sometimes, the delivered speech may contradict with official regulations.

Cross-border conflicts

The management of cross-border natural resources with diverse and contrasting objectives has the potential to result in conflicts among diverse trans-boundary actors. A part of Lake Jipe falls within Tsavo National Park on the Kenyan side. Fishing is prohibited in this area. Nevertheless, fishers from the Tanzanian side enter the park area illicitly. They illicitly fish in this area for the same reason the area is protected, because of the abundance of big fish. When caught, besides being accused of illegally entering the park, fishers are also accused of illegally crossing the border and carrying out fishing in the prohibited area. Claims of entering the country without a legal permit, however, are rarely raised when this occurs. In normal circumstances, this is not a pertinent violation because inhabitants of the Lake Jipe area (Tanzanians and Kenyans) intermarry and intermix in social and economic functions without having legal permits to cross the border. It is possible to find Tanzanian women selling fish on the local markets in Kenyan side, or Kenyan vendors selling vegetables and buying milk on the Tanzanian side. Leaders at the international border cooperate in the resolution of socioeconomic matters involving their people regardless of their legal national identity. Nonetheless, in the case of fishing in the park area, the issue of legal identities becomes pertinent. Settlement of this kind of problem requires adhering to legal procedures (Table 4.7), which necessary involves their countries of origin.

Table 4.7 shows some fishers from Tanzania who were fined and/or sentenced to jail in Kenya because of illicit fishing in the protected area of Lake Jipe on the Kenyan side, which is within the Tsavo National Park. Besides arresting and imprisoning the criminal fishers, complaints against fishers from Tanzania encroaching on protected areas in Kenya are also raised in letters to Tanzanian authorities (e.g. the Ward Councillor, Ward Executive Officer, etc.). Further, the violators sometimes request, through their relatives who visit them while in captivity, that the authorities in Tanzania send formal requests to Kenyan authorities to allow them serve their punishment in Tanzania.

Although the fishers know that it is forbidden to enter and to fish in the Tsavo National Park, and are aware of the consequences they are likely to encounter when caught, they chance fishing in the protected area, with the awareness that, if not caught, they will obtain substantial amounts of big fish that will enhance their incomes. This is contrary to the thinking of the Natural Resources Officers at the Mwanga district, who believes it is possible to eliminate this cross border conflict by educating the fishers as to the negative impacts of not observing the international boundaries.

Table 4.7. Actions taken against Tanzanian fishers who trespass into protected areas in Kenya on 24 May, 2005 (source: Natural Resources Office, Mwanga district, 2007).

Violators	Action taken
Fisher 1	6 months imprisonment
Fisher 2	6 months imprisonment
Fisher 3	6 months imprisonment
Fisher 4	Fined TZS 175,200

4.4.2 Conflict resolution

In this sub-section, I analyse mechanisms used to address conflicts emergent in the use and management of fisheries resources. We have already seen that conflicts occur at the community level (involving the fishers, fisheries committee, and the village, ward and division governmental levels) and at the district level between the community (fisheries committee) and the government. Mechanisms used to resolve these conflicts depend on the nature of the conflict, the degree of the conflict, the level at which it occurs, and the actors involved.

Collaboration between the government (formal institutions) and the community (informal institutions) is one way of resolving conflicts in fisheries management. An example of this collaboration is that which integrates the elders in the community and the government at the village and district levels when conflicts exceed the power of the community to resolve. Most conflicts involving residents are resolved by elders without involving the government. In addressing these disputes (among the residents), elders employ traditions and customs that are legitimate among the residents. These may include fining offenders, or preparing a meal and local brew and eating together while mediating the dispute. This mechanism is based on the legitimate authority and held by elders in the community.

However, when conflicts involve both outsiders (non-residents) and residents, the outsiders do not recognise the authority and legitimacy of the traditional arrangements. Although the elders can ease the tension between users to a certain extent (especially by restraining these residents from reacting violently), they need to cooperate with the government to address this kind of conflict. In such conflicts, the use of weapons, such as knives, spears and arrows, may occur, as took place at Mkisha subvillage at Lake Jipe. In such situations, the elders inform the District Natural Resources Office and the Police. The District Natural Resources Office visits the area and collaborates with the Divisional Secretary and elders to resolve the conflict by advising fishers to respect one another's fishing sites. The level of conflict thus necessitates the collaboration of the community and various governmental levels. Whereas the elders are the first to intervene in this dispute, the level of the conflict threatens the elders' ability to manage the situation. This causes the elders to call for the involvement of other authoritative bodies.

Formal bureaucratic procedures involve consultative meetings between the District Commissioner (DC) and the fishing community. In these meetings, the DC is accompanied by the District Natural Resources Officer, who strategises with the fishing communities about mechanisms for combating illicit fishing. The DC's responded to the accusations levelled by the fisheries committee against the District Natural Resources Officer through a consultative meeting.

The use of established legal mechanisms is imperative for resolving conflicts due to violation of the by-laws. This is especially the case when some actors physically harm others when told to stop using unsustainable natural resource exploitation methods and equipment. Conflicts that result in harm to other actors are addressed in the Fisheries Act 22 of 2003 and the district fisheries by-laws. For example, Article 41 of the act stipulates that anyone who obstructs the entrusted authorities from exercising their jurisdictions commits an offence and is liable, upon conviction, to be imprisoned for a period not exceeding six years.

Integrated legal and diplomatic mechanisms are used to mediate cross-border conflicts at Lake Jipe. Encroachment on the Tsavo National Park situated on the Kenyan side of Lake Jipe by fishers from the Tanzanian side requires collaboration between leaders from the two sides to resolve emerging disputes. Kenyan ward leaders communicate with ward leaders in Tanzania about the illicit fishing practises of Tanzanian fishers. The Ward Councillor reports this to the District Natural Resources Officer for action. Besides such collaboration, legal actions are taken against illicit fishers, including fines and imprisonment, as was indicated in Table 4.7. The two countries may negotiate as to whether the fishers should be imprisoned in Kenya or should be returned to Tanzania and serve their punishments in Tanzanian prison.

Local stakeholders' meetings are another approach to conflict resolution, though the success of such an approach is not always certain. Participatory approaches are used to address conflicts between village leaders and non-resident fishers, who do not observe the regulations and threaten the leaders at Lake Jipe. A village assembly is convened in which all villagers, including resident and non-resident fishers, are entitled to participate. Leaders at the ward and village levels explain the established procedures and encourage non-resident fishers to observe them and to stop threatening the leaders. This method has to be implemented on a regular basis because non-resident fishers enter and leave the area frequently.

Since this thesis investigates the possibility of co-management of natural resources between the government and local people, the next section identifies and evaluates co-management arrangements in the management of fisheries resources at Lake Jipe.

4.6 Co-management analysis and evaluation

Co-management exists in fisheries management through the interplay of formal and informal institutions in fisheries management at Lake Jipe. Formal institutions are government-based institutions. While governmental actors enforce some formal institutions, others formal institutions have been entrusted to the community and community organisations steer their actions, while the government facilitate those actions. The fisheries committee is one such organisation, which is appointed by the community to enforce fisheries policy through monitoring, control and surveillance, of natural resources use practises in the users' community. These responsibilities have been entrusted to the community by the government, and are implemented within the framework of and with the help of government institutions. The government retains the jurisdictions of collecting revenues from taxes, fishing and fishing vessels licenses. At the community level, there are also informal institutions, which are based on norms, traditions, and customs, that evolve to resolve socioeconomic and cultural issues emerging in interactions between users of fisheries resources. These interact with the formal institutions to govern and mediate fisheries resource use and management among users.

Fishers are important actors in the co-management of fisheries resources at Lake Jipe. However, co-management involves resident fishers, while non-resident fishers are excluded because they are viewed as a threat to fisheries resource management. They are commonly identified as the perpetrators of illicit fishing practises. It is, nevertheless, questionable as to whether excluding non-resident fishers is a solution to their unsustainable practises. Because they are excluded, non-resident fishers may regard themselves as not having a stake in the resource, and, therefore, may continue to perpetuate unsustainable resource use practises. However, while the inclusion of non-resident fishers in management committees may help to engender a sense of ownership over resource management, it is not clear how they should be integrated. This is because non-resident fishers do not stay permanently in one area. While today they may be at Lake Jipe, tomorrow they will be at another area, and they may return to Lake Jipe at a later time. Moreover, it is probable that new non-resident fishers are likely to come to Lake Jipe area in different fishing seasons. While the exclusion of non-resident fishers may not provide a solution to unsustainable fisheries practises, their inclusion needs to be considered with care in order to accommodate their tendency to migrate seasonally.

The participation of governmental actors at the local level (village, ward, division) is contradictory. While in some areas, some local leaders seem to support co-management initiatives, in other areas they seem to block or frustrate the co-management arrangement.

In some cases, they even seem to align with violators of fisheries regulations and institutions as is implied not only in letters from fisheries institutions to the district leaders, but also in letters from the Ward Fisheries Officers to the district leaders. It seems that some local government leaders do not support co-management arrangements because they benefit from illicit collaborations. There are therefore divisions in the implementation of fisheries management among the local leaders, and between the local leaders and the district leaders. While the district leaders generally promote the implementation and enforcement of fisheries regulations, the institutions at the local levels (village, ward, and division) are not well-integrated. Some officers support the enforcement of fisheries policy and have cooperated well with fisheries committees, to the extent that one committee became concerned with an officer's working constraint (i.e. the lack of transportation) and tried to alleviate it. In contrast, other officers obstruct the enforcement of fisheries regulations, and devise mechanisms to acquire illegal revenues. Instead of facilitating the committees, they constrain them, and do not seem to cooperate even when the committee communicates with them. For these officers, the initiation of the fisheries committees may appear as a threat to their power, and to the benefits they acquired before the establishment of these institutions. These mixed interactions contribute to either the success or the weakening of the co-management arrangements.

The frequent involvement of the police in co-management may imply that co-management is not supported by the fishers and community users at Lake Jipe. It appears that the fishing community is not ready for co-management, but that co-management has been thrust upon them. When the police are frequently involved in the policing of the co-management arrangements, one can ask whether these resource management regimes are sustainable. Similarly, the low level of cooperation between the fishing community and the fisheries committee, and the prevailing conflicts between the fishers and the fisheries committee, may raise the argument that the co-management arrangement currently prevailing in fisheries management at Lake Jipe is an imposition, rather than a legitimate and widely accepted arrangement. While the fisheries committees are entrusted with the responsibility of enforcing compliance to fisheries policy in the fishing community, their work need to be legitimated in the eyes of the fishing community who are the target of fisheries management initiatives. The frequent police coercion over the local users to observe fisheries regulations, and the low cooperation received by community actors carrying out enforcement responsibilities, may imply that the co-management regime will not be sustainable in the long term.

Informal institutions at Lake Jipe have a role to play in the co-management arrangements, and these collaborate with formal institutions. The informal institutions are norms and customs that govern relationships among specific resource users. At Lake Jipe, social norms and traditions give elders the power to mediate conflicts among resource users. Although the local or district government is usually ultimately responsible for the resolution of conflicts between resident and non-resident users of fisheries resources, the elders assist these actors in resolving conflicts, especially at the onset stage, because these actors are close to the people. In most cases, the elders report these conflicts to the government leaders because, even at the village level, there may be some delays inherent in the channels that relay information to governmental leaders. Moreover, at other times, at the village level, the government may fail to suppress the conflicts before higher level governmental actors are informed. In such cases, the elders may manage to calm the fishers before these governmental actors arrive at the scene, as was the case in Mkisha sub-village. Their power to resolve these conflicts is due to the legitimacy of elders in the sight of the local people, including the resident fishers. However, the elders' ability to resolve conflicts is higher within the community than in a situation that comprises external and internal users. This is because the external users (non-resident fishers) do not recognise the elders' legitimacy as traditional leaders.

One of the most important actors implementing fisheries co-management between the government and the community at Lake Jipe is the fisheries committees. This type of committee can be seen as a co-management arrangement in itself, as the committees are given resources from the government but are appointed by the community. The fisheries committees regard the responsibilities entrusted to them seriously. They even challenge the District Government Natural Resources Department when they perceive this authority to be disabling their efforts to curb illicit fisheries practises, as is indicated in the example in section 4.4.1. However, as mentioned earlier, the committees are not accorded due cooperation by the fishing communities, who view them as an imposition from the government, rather than as representatives of the community that appointed the committee members. As we have also seen, the fisheries committees know where to report to in case a governmental level is not responsive. This knowledge allows the committees to bypass non-responsive governmental levels and actors, and to report unaccountable leaders to a higher institutional level. These committees are composed of resident fishers, and therefore there is a general tension between the committee members and non-resident fishers. There is a belief that if non-resident fishers were included in the fisheries committees, they would be likely to misuse their position to perpetuate unsustainable fishing practises. For fisheries committees to sustain their efforts, however, they need support from the government and other community actors in their areas to create a smooth management process rather than escalating conflicts among actors at the local level.

The Fisheries Section (in the Natural Resource Department) at the district level is an important formal actor for overseeing formal fisheries management institutions by enforcing, facilitating, and coordinating collaborative fisheries management at the district level. It is the enforcer of the policy and legal instruments at the local level. However, it may appear that the District Fisheries Section focuses more on procuring revenues from local resource areas than building the capacity of fishers. For example, the provision of training to the fishers, the exploiters of fisheries resources, is not prioritised. If sustainable management of the fisheries resources is to be attained, training should be given, not to the fisheries committee alone, but also to fishers. The fishers need to gain skills on sustainable fishing practises and methods. The Fisheries Section currently excuses this lack of training by saying that fishers would not alter their behaviour, even if trained. This point is disputable. Similarly, although non-resident fishers are excluded in the management of fisheries resources, the district department has not made any effort to negotiate participatory fisheries management among these users.

The existing co-management arrangement, therefore, is composed of formal and informal institutions. Collaboration between these institutions (through government by-laws and regulations and the norms and customs of the community as guided by the elders) enables the resolution of conflicts among fisheries resource users and those responsible for fisheries management. While formal (government-based) institutions alone mainly resolve fisheries management issues involving governmental actors and the community based fisheries committee, and between the institutional enforcers and the users, informal institutions mainly resolve issues emerging among resident resource users. However, these formal (state) and informal (community based) institutions integrate or cooperate to resolve resource use conflicts that involve both external and local users, which cannot be adequately resolved by formal or informal institutions individually. Although community informal institutions have not been recognised and entrusted by the government to partake in the management of fisheries resources, they have joined the process based on their authority in the community. The power of the local institutions, however, is limited when resource use issues comprise local and external users. In these cases, the integration and collaboration of formal and informal (local institutions) is necessary.

While the fisheries co-management arrangement works to resolve resource use conflicts between resident and non-resident fishers and between fishers and enforcers of sustainable fisheries management, it focuses on the fisheries sector alone. Other sectors that influence the management of fisheries resources at Lake Jipe (e.g. livestock and agriculture) are ignored. Suffice it to mention here that the problem of siltation of Lake Jipe is contributed to by, among other things, unsustainable farming practises downstream and upstream of Lake Jipe. The management of Lake Jipe resources is, therefore, a multi- and cross-sectoral issue. The next chapter will present an analysis the collaboration between the government, non-governmental actors, and the community in the management of natural resources in agricultural production.

Chapter 5.
Management of natural resources in agricultural production at Lake Jipe

5.1 Introduction

The sustainable management of Lake Jipe resources in agricultural production depends on, among other things, collaboration between the government, non-governmental actors, and agricultural producers. The users of natural resources in agricultural production at Lake Jipe comprise upstream and downstream farmers. The two localities are interconnected by water that flows from the upstream area to the downstream area. Therefore, poor water management practises upstream have negative consequences for water downstream. Land also connects the two areas, so that land management practises at the upstream area have implications for the downstream area. Poor land management at the upstream area results in soil erosion and deposition to the bed of Lake Jipe, which is located downstream. Whereas the farmers are the primary managers of natural resources at Lake Jipe, governmental and non-governmental actors are the facilitators and enforcers of good management practises.

This chapter analyses co-management arrangements between the government (agricultural sector) and the community (farmers) for sustainable management of Lake Jipe. Specifically, the chapter analyses how governmental, non-governmental and community institutions collaborate in managing land and water resources in agricultural production. The paper begins with a brief overview of environmental problems in the management of natural resources in agricultural production at Lake Jipe. The chapter then analyses how governmental, non-governmental, and community institutions participate in resolving conflicts that emerge in resource management. Finally, it analyses and evaluates co-management between governmental, non-governmental and community actors and states the implications for Lake Jipe.

5.2 Environmental problems in Lake Jipe due to agriculture

The main environmental problem at Lake Jipe is siltation caused by inappropriate agricultural activities. Two major sources of the problem are land degradation on steep slopes and the encroachment of cultivation on fragile areas such as river tributaries and the shores of Lake Jipe.

Inappropriate cultivation practises degrade land resources. Cultivation methods used at Lake Jipe include flat cultivation, fallow, and the use of ridges, terraces and contours. Flat cultivation is a land tillage method whereby a hand-hoe or plough is used to cut and turn the soil, and usually stalks are burned. When it is practised on steep areas, it results in soil erosion due to a lack of soil harvesting structures. Fallow is the practise whereby a farm plot is left uncultivated for a specified period in an attempt to replenish its potential for future productivity. The flat cultivation method is the main farming system practised by farmers at the Lake Jipe area, even on steep areas. For example, out of 80 farmers interviewed, 49.2% practise flat cultivation, 24% practise fallow, 19.5% use ridges in their farms, 3.8% use terraces, and 3.8% use contours (Figure 5.1). This implies that the majority of the farmers have not adopted soil conservation techniques, such as terracing or contours. Although they make use of crop rotation and inter-cropping techniques as methods to improve soil fertility, their farms remain vulnerable to soil erosion caused by sloping terrain, especially during rains. The Lake Jipe Ward Extension Officer noted that the farmers' failure to adopt these techniques may result from the fact that contours and terraces are labour-intensive and that they require relatively high cash investments, which farmers cannot afford.

Soil erosion leads to a decline in production due to a decrease in soil fertility over time. Although it was not possible to get statistical data to substantiate this argument, interviews with farmers indicate that agricultural production has been in decline over recent years (Table 5.1). When their farms are less productive, farmers may move to fragile areas because arable land is limited. This encroachment on fragile areas, which farmers' view as relatively fertile areas, may result in the development of gully erosion, as indicated in Figure 5.2. The eroded soils are deposited on the bed of Lake Jipe.

Even though farming is prohibited on the steep mountainous areas inclining toward Lake Jipe, these areas are encroached upon by cultivation practises. Moreover, the farming techniques carried out on these fragile areas are poor. These encroached-upon areas are located both downstream and upstream of Lake Jipe.

Poor enforcement of conservation by-laws is a contributing factor to the encroachment upon these areas. They are public lands that are managed by the Natural Resources Department of Mwanga district. Monitoring of these areas is ineffective, however, creating opportunities for

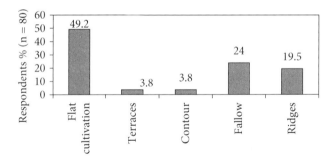

Figure 5.1. Management systems used in agricultural production at Lake Jipe.

Table 5.1. Consequences of soil erosion at Lake Jipe (source: field survey, 2007).

Consequences	Respondents (%); n=80)
Less harvest	65
Soil infertility	40
Abandonment of some farm plots	30

Figure 5.2. A formerly cultivated area in the proximity of Lake Jipe is now only dry gullies due to rain erosion. Shrubby vegetation and water-weeds at the Lake Jipe are visible in the background.

encroachment by human activities. These encroached-upon areas are quickly degraded, and the soils they once contained erode and are washed towards Lake Jipe.

The appropriation of water resources is another problem in natural resource management at Lake Jipe. Agricultural activities around the water resources destroy vegetation on the shore. This occurs not only close to the shores of Lake Jipe but also near tributary rivers. This problem escalates during dry seasons. Farmers encroach upon areas close to the water resources in an attempt to enhance food security and reduce the likelihood of food shortages. They are aware that the government prohibits cultivation in the fragile areas (e.g. close to water sources), but their view is that government regulations are difficult to enforce during such a time. Since the vegetation close to the encroached-upon water resources act as a filter for and barrier to silt that is carried by water that flows from the upland areas, the degraded areas enhance the rate of silt deposition into Lake Jipe and its water sources.

5.3 Land management practises

In analysing sustainable management at Lake Jipe, the land use practises of upstream and downstream users merit consideration. As we have seen in the preceding section, the siltation problem at Lake Jipe is caused by inappropriate farming practises at both downstream and upstream areas. Both areas are therefore important to understand because even when farming practises on the downstream are sound, if farming practises on the upstream are poor, sustainable management of the lake will not be attained. This section therefore analyses farming practises of both localities, the areas upstream and downstream of Lake Jipe.

5.3.1 Farming practises

Common farming practises at the areas upstream and downstream of Lake Jipe are inter-cropping, agroforestry, crop rotation, flat cultivation, ridges, contours, and terraces (Figure 5.3). Inter-cropping is an intensive farming technique whereby various crops are combined on a single unit of land. Crops used in this kind of farming are usually beans and maize. Farmers utilise this practise due to a shortage of land available to practise monocropping, and because it is a strategy of spreading risk through diversification.

Crop rotation is a practise designed to increase soil fertility by alternating crops between planting seasons. Again, the two crops commonly grown under this practice are beans and maize. Maize production increases when maize is cultivated on a plot that was previously used for beans. Since land is a limiting factor, the decision to practise crop rotation or inter-cropping depends on the farmer's intention to either increase the yield of one crop or to spread the risk among different crops in a single season.

Agroforestry combines trees and crops on the same land management unit. The agroforestry system is useful in controlling soil erosion on steep slopes, and can minimize soil and wind erosion in lowland areas. This system provides the dual benefits of conserving the environment and providing socioeconomic benefits, such as timber, firewood, and fruit, to people.

Flat cultivation is common in some sloping land and valley floors where food crops are cultivated. Whereas maize and beans are planted on sloping areas, valley floors are used to grow maize, beans, sugarcane, banana and yams. Land is vulnerable under these management systems because cultivation on steep slopes does not use appropriate techniques for soil erosion control. Further, in the valley floors, the practises are extended to edges of water resources. The cultivation of crops such as banana, sugarcane and cocoyam at or near water sources and river systems may reduce water flow downstream.

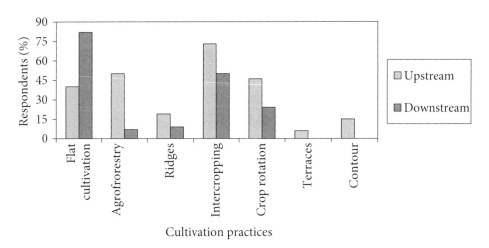

Figure 5.3. Farming practises of the farmers located upstream and downstream of the Lake Jipe area.

Some practises, such as the use of ridges, contours, and terraces are generally practised on a small scale. These practises have not been common in the community but are the result of interventions by government extension workers and non-governmental agents, such as MIFIPRO. These have been introduced as a way of addressing land degradation problems due to poor land husbandry systems. Since land is limited, some practises, such as contours, are meant to provide multiple land use benefits to farmers, such as fodder for livestock and controlling soil erosion by adopting an intensive management system. Ridges and terraces are potentially important for controlling soil erosion and conserving soil moisture.

Various approaches have been applied to promote these techniques, including the use of farmer field schools. In these schools, farmers learn through demonstrations and practise various methods of managing soil and improving land productivity. Adoption of these technologies, however, is low due to a variety of factors (Table 5.2). A shortage of land is a constraint to the adoption of contour use. Nearly three quarters (72%) of farmers acknowledge the importance of contours in providing the dual benefits of livestock fodder and controlling soil erosion, but opt to plant fodder on land that cannot be used for agricultural production due to a shortage of land for food crop cultivation. Constraints to adopting ridges and terraces include labour shortages (37.5%), and a lack of funds to invest in these (62.5%). Farmers stated that they are incapable of investing in these structures due to the deterioration of their income caused by low agricultural productivity. There are therefore many challenges to the economic and environmental feasibility of these important but under-utilised agricultural techniques.

In terms of natural resource management, the upstream and downstream areas are connected. Unsustainable land use practises upstream lead to the movement and deposition of eroded soils into water streams, causing pollution downstream. Similarly, upstream farming practises close to water sources degrade those sources, leading to a decline in water flow. The farmers interviewed in this study have noted a decrease in water flow in rivers, and they attribute this situation to land degradation (21.7%), drought (17.4%), a lack of adequate rainfall (17.4%), and degradation of indigenous trees (4.3%) (Figure 5.4). Although indigenous trees are crucial for water conservation, there is only a low level of awareness of their importance among farmers.

Table 5.2. Reasons for under- or non-adoption of sustainable land use techniques (Source: field research, 2007).

Reason	Respondents (%) n=80
Lack of funds to invest in techniques	62.5%
Lack of labour	37.5%
Land shortage	72%

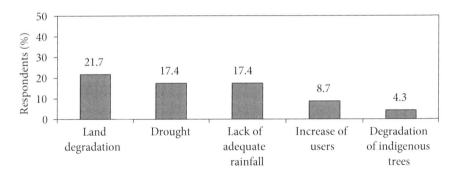

Figure 5.4. Farmers' reasons for the decline of water flow at the upstream area (n=80).

5.3.2 Capacity building in farming practises

The management of land in agricultural production at Lake Jipe is related to capacity-building of farmers, which is performed by a non-governmental agency (Mixed Farming Improvement Project, (MIFIPRO)) and the government. Since poor cultivation practises upstream result in soil erosion and the deposition of eroded soil into downstream Lake Jipe, individuals practising farming on sloping areas are trained in sustainable land management practises by these groups. Farming techniques are taught through training seminars and participatory experiential learning techniques, such as farmers field school (FFS). In these educational programmes, farmers practice various techniques, such as ridges, contours.

These two groups also offer farmers education in the form of advice. Since the Lake Jipe area, especially the lowland region, is commonly stricken by drought, government technicians advise farmers to grow drought resistant crops, such as cassava, cow-peas, lima beans, and sorghum for food purposes, and cotton as a cash crop. To motivate the farmers to adopt these technologies, the government provides farmers with seeds of these crops free of charge. Since these crops can thrive in drought conditions, their adoption has implications on reducing the exploitation of fragile areas close to the river margins and along the shore of Lake Jipe.

However, the rate of adoption of these technologies by farmers is low. While some farmers claim that the techniques (i.e. contours, ridges, terraces, and agroforestry) are labour and cost intensive (Lake Jipe Ward Agricultural Officer 2008), other technologies that do not require such investments are also not adopted due to a lack of preference. In some cases, educational staff have limited knowledge as to farmers' socioeconomic and socio-cultural interests before introducing these interventions. For example, during the drought season of 2006, when there was a severe food shortage at Lake Jipe, the government, through its technical staff, gave the farmers sorghum seeds to plant for food production. However, the *Pare* community regards sorghum as food for animals, not for human consumption. If one eats sorghum, the rest of the community will look down upon that person as the poorest in the community. The seeds, therefore, remained untouched by the farmers. The government had only a limited understanding of the farmers' socio-cultural setting before advising them to sow sorghum, and consequently farmers did not adopt the advice[2]. Therefore, two-way communication is imperative for putting participatory natural resource management into practise.

5.4 Water management practises

In this section, the irrigation practises of farmers at Lake Jipe will be analysed and the capacity-building arrangements and practises facilitated by governmental and non-governmental actors will be discussed.

Water is used for agricultural production under irrigation farming. Irrigation farming is conducted both upstream and downstream of Lake Jipe. In the upstream area, it is conducted on valley floors. Except for some farmers that learned border irrigation from their counterparts in the downstream areas, traditional surface irrigation practises are generally implemented upstream. Beans, maize and vegetables such as cabbage, tomatoes and spinach are cultivated on these valley floors. The people in the upstream areas must cope with a shortage of arable land. Nevertheless, the use of valleys for irrigation farming is not environmentally sustainable. Because of the land shortage and a high food demand from the upstream population, the irrigation practises have come to encroach upon local water resources. This may reduce water flow to the downstream areas.

Irrigation practises in the downstream area depend on water flowing from the upstream areas. In contrast to the upstream areas, where farming practises are carried out on an individual basis, the farmers downstream have formed a farmers' organisation. This organisation was formed to manage water allocation issues. The organisation bargains over water allocation with upstream farmers.

The main irrigation practises on the downstream are the traditional surface flood irrigation, and the improved irrigation practises using the border technique. Farmers have practised traditional surface flood irrigation in the region from time immemorial. However, field observations and interviews with farmers (Figure 5.5) reveal that this irrigation method wastes water, and creates soil erosion through surface run-off. Further, this method is time-intensive, especially where water is scarce. Some farmers work until midnight irrigating their fields due to the amount of time required for irrigation. This irrigation system wastes water through percolation as it flows

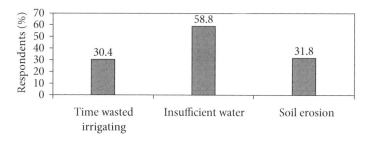

Figure 5.5. Problems encountered in irrigation farming at Lake Jipe (n=110).

_____

[2] This was revealed by a farmer at Lake Jipe.

from the intake at the parent river, along the furrow, and finally to the fields. Canal erosion slows water flow rate, so that less water arrives at the fields than is taken from the water source.

Another irrigation practise is improved irrigation using borders. With this system, an irrigation furrow is constructed from the parent river to the fields with bricks and cement. A farm plot is split into subplots with raised borders that concentrate water in one plot until it is wet enough, and then the water is allowed through an outlet to another plot. The flow rate is high with this method. There is reduced soil erosion because the surface run-off speed is reduced, and less time is required for irrigation (Figure 5.6). While farmers with traditional furrows may irrigate until midnight, those using improved furrows only have to irrigate until 6:00 pm.

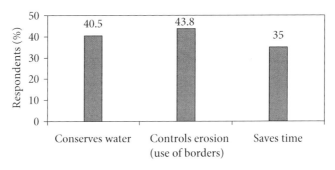

Figure 5.6. Advantages of the improved irrigation furrow at Lake Jipe (n=80).

5.4.2 Capacity building in irrigation practises

The improved irrigation technology has been introduced by the Mixed Farming Improvement Project (MIFIPRO). In addition to providing education about the use of borders in irrigation, MIFIPRO facilitated farmers' construction of improved irrigation infrastructure for some furrows. MIFIPRO mobilised farmers to contribute to the purchase of construction materials, provide their labour and contribute to the training staff's transportation costs as part of the construction project. The farmers who provided their labour and contributed money became the owners of the improved furrows. While irrigation knowledge is useful, it was not provided to all farmers in the area. The farmers who could not contribute were excluded, and consequently conflicts arose between the farmers included in the project and those who were excluded.

Since the improved irrigation structures save time, some farmers still operating under the traditional system secretly used the improved irrigation structures at night to irrigate their fields. They irrigated secretly because the improved irrigation infrastructure was owned by the farmers that contributed their resources to its construction. Although water use was the right of any villager, farmers had to pay a membership fee to use the improved furrow. This membership fee for one season was 50,000 Tanzanian shillings (TZS) (€ 27.80). The farmers that owned the furrow used this method to exclude other farmers, and thus to control the number of farmers that could be served by the improved structure. Though it produced important benefits, the MIFIPRO intervention therefore also introduced conflicts to the farming community between those who benefited from the intervention and those who did not.

While the original irrigation border intervention did not assist all farmers, some farmers have nevertheless learned border techniques from those that did directly benefit. For example, farmers that carry out irrigation farming downstream but reside upstream have disseminated knowledge about border irrigation to the upstream areas, and some farmers already practise it.

The government is a partner of MIFIPRO in the provision of irrigation services (Figure 5.7) and in recent years has come to build on MIFIPRO's work through an additional intervention. This intervention is made possible through agricultural development funds from the Participatory Agricultural Development and Empowerment Project (PADEP). Farmers are encouraged to choose projects they want to implement and are trained on project proposal writing with assistance from agricultural technical staff and the Ward Executive Officer. This is followed by an assessment of the potential environmental impacts of the projects by a District Facilitation Team (DFT). Should technical problems arise, farmers are advised to revise their proposals. The resulting project proposals are forwarded to the PADEP headquarters at the Ministry of Agriculture, Food Security and Cooperatives for funding. Funds then flow back from the Ministry to the farmers along the same chain. Irrigation projects were the type most commonly proposed by farmers at Lake Jipe, and therefore seem to be their priority need.

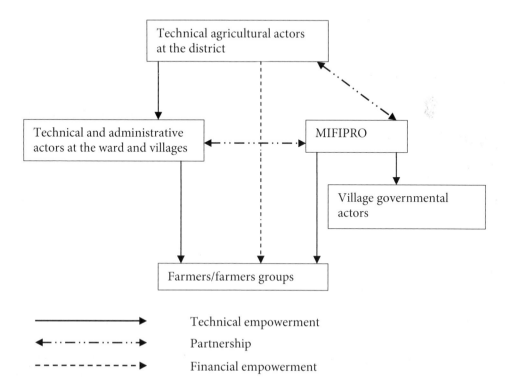

Figure 5.7. Partnership between the government and MIFIPRO in capacity building of the farmers.

In contrast to MIFIPRO's intervention, less time is spent in the community in implementing the government's intervention. Although some farmers using traditional irrigation management practises have been offered irrigation infrastructure through the Participatory Agricultural Development and Empowerment Project (PADEP), less time is invested in irrigation practises because the government assumes that farmers have already been educated about these practises by MIFIPRO. This presumption may be incorrect, however. In some areas, especially those not yet accessed by MIFIPRO, farmers still use traditional irrigation techniques instead of the border technology. In such areas, both training on irrigation practises and improved irrigation technology is needed.

As with MIFIPRO, before the government grants funds to farmers, farmers must pay 20% of the project costs. This requirement is based on the assumption that it will create a sense of ownership over the project. The government also feels that it cannot offer everything to farmers, but that agricultural development must be a partnership between the government and the farmers. In acknowledgement of this, the government gives farmers some freedom of choice in irrigation infrastructure development. Farmers decide what contractor to engage in the construction of irrigation infrastructure. Tendering approaches are adopted, whereby potential contractors can apply for construction work. This is intended to empower farmers to make their own decisions. These practises are indicative of shared responsibilities in the management water in agricultural production, where the administration is shared between the government and farmers. Instead of the government dictating who should construct irrigation infrastructures, farmers are provided with profiles and addresses of potential contractors, and they decide who should be involved in the construction.

To summarise the foregoing discussion, in the upstream-downstream relationship in agricultural use of natural resources, three aspects are important. First, knowledge flows between the two areas. The people downstream have learned farming practises from the upstream areas, which are their areas of origin, and they generally perpetuate these practices. Further, the new techniques learned downstream through initiatives are transferred by individual farmers to the upstream area. Second, water links the two areas. The people downstream are dependent on water that flows from the upstream area. Third, the two areas are connected by the movement of soil caused by poor farming practises upstream. Though the movement of soil has little impact on the upstream users, it effects the downstream farmers negatively. Co-management arrangements must therefore consider both localities to be successful at promoting sustainability.

Governmental and non-governmental actors play important roles in building the capacity of farmers in both localities for sustainable management of natural resources in agricultural production. A partnership exists between the government and the MIFIPRO in the provision of financial and technical support for water and land management. This partnership has not been without its problems, however, as some constraints create conflicts or impair the adoptability of some technologies. For example, the introduction of the irrigation services among some farmers created conflicts between those farmers and others who did not have legitimate access to the improved structures. Other constraints are related to farmers' financial inability to put available technologies to work, and poor understandings of farmers' socio-cultural setting, resulting in some sustainable recommendations not being adopted.

As outlined in Chapter 2, natural resource management in Tanzania involves arrangements, formal and informal institutions, and governmental, non-governmental, and community actors. Institutions govern the practises of the actors to attain sustainable management of natural resources while ensuring the procurement of socioeconomic benefits. In the coming section, the institutions governing the management of natural resources in agricultural production in the Lake Jipe area will be analysed.

5.5 Institutions relevant to natural resource management in agricultural production and conflict resolution at Lake Jipe

In this section, I identify and discuss committees, institutions and regulations that are relevant to the use of natural resources in agricultural production and conflict management at Lake Jipe. Both the government and community institutions, including farmers' institutions participate in enforcing the sustainable management of natural resources in agricultural production. Relevant formal and informal institutions and regulations include forest conservation by-laws, village environmental by-laws, improved land use regulations, water management institutions, and customary regulations. While some of these institutions and regulations are governed through cooperation between the community/farmers and the government (e.g. the environmental by-laws), others are entrusted to farmers alone, and others are governed by local custom.

5.5.1 Ward/Village Land Councils

Land has multiple uses and involves complex tenure arrangements and processes that require mechanisms for governing sustainable use and mediating conflicts of interests among various stakeholders. At Lake Jipe, land councils were formed in the early 2000s for the purpose of administering and mediating land use and management. Their formation was in line with Village Land Act of 1999. Although the formation of these councils was top-down from the district to the ward and village levels, conflicts in land use among various users within and between different sectors at the village level was the basis for the decision to create the district land councils. The appointment of land council members is a participatory process that involves the village community through the village assembly. The councils consist of villagers who are not members of the village government. A pre-requisite for their appointment is that they have to have knowledge of the village area and its boundaries.

Land councils address land issues occurring in formal government-allocated land and in customary land tenure arrangements. On land allocated under the governmental system, formal land laws apply, whereas for land allocated and managed under the customary system, customary rules apply. Land councils, therefore, employ both formal and informal rules in dealing with issues related to land administration, use and management in villages at Lake Jipe. For land that is managed under the customary system, decision-making is mainly based on customary institutions, though formal land institutions enforce compliance to customary institutions when the need arises.

While land councils at the village level consist of villagers and elders who oversee customary regulations in their clans, the ward land council comprises elders that represent village interests and ward-level governmental leaders, who represent the ward development committees (WDC).

The ward land council is therefore a co-management institution and uses a combination of formal regulations and customary rules. When land-use issues cannot be resolved by the village council, they are presented to the ward council, which uses both customary and formal rules collaborate to resolve the issues. If these local councils fail to resolve land disputes, those disputes are then addressed at the district level.

5.5.2 Ward tribunal

The ward tribunal acts as a local judiciary. It is the lowest body in the organisational structure of the Tanzanian judicial system. This multidisciplinary institution governs various socioeconomic and environmental issues among the local people in their villages. This institution deals with issues that include those pertaining to land use and conflicts among land users. However, in order for issues related to land use to be presented to the ward tribunal, they must first have been brought to the village and ward land councils, where they were not able to be resolved. This institution also operates under customary rules if land disputes occur in land that is under customary land management. If a case fails to be addressed at this level, it is forwarded to the district land council and to court for further mediation and arbitration.

5.5.3 Environmental by-laws

Environmental by-laws at the village level are responsible for controlling environmental degradation. They enforce sustainable environmental conservation by prohibiting cultivation practises around water resources, close to the lake, on steep areas, and in forest areas under village jurisdiction. Administration and enforcement of the by-laws is entrusted to a committee appointed through the village assembly and led by members not included in government leadership (e.g. the chairperson and the secretary of the committee are community members). Nonetheless, though they may not assume leadership positions, some members of the village government are members of the committee. At Jipe village, for example, of the ten village environmental committee members, four are from the village government (sub-village leaders).

The district government, through the District Natural Resources Department, influences the formation of an environmental institution, and a district legal officer facilitates the process technically. When environmental by-laws were first established at Lake Jipe, a small committee, comprising the village government and the district legal officer, prepared draft by-laws. Subsequently, the village assembly was convened and the community was informed of the regulations, and they were given 30 days to give comments. After the 30 days, the by-laws were approved and enforcement began. The village assembly also appointed members of the environmental committee from the community. Committee members from village government were not selected but were integrated by virtue of their positions as leaders in their sub-villages.

The village environmental committee initially came into being with no capacity to enforce by-laws. Although the district influenced the formation of this institution, it did little to build the committee's capacity for enforcement. As member of the committee noted:

It was easy for the government to mobilise the formation of environmental committees, but it forgot to empower the committee after its formation; we don't know how to properly perform our responsibilities. The committee is physically present but functionally incapable (Hatibu, 23/7/2007).

I uncovered through the interview that there were no mechanisms for evaluating the status of environmental degradation in the village. Realising this weakness, MIFIPRO has assisted in the empowerment of this committee through training and explaining the environmental by-laws. However, because their responsibilities are multifaceted, significantly more training is required to empower the committee.

5.5.4 *Water management committees*

Two kinds of water management committees exist at Lake Jipe. These are both formal committees, and they are the village water committee and the lake committee. The village water committee was formed as a requirement of governmental water law, whereas the formation of the lake committee was influenced by MIFIPRO. The village water committee enforces the management of water for domestic consumption in villages. Through the village assembly, villagers appoint village water committees. Its members are villagers who are not leaders in village governments. The lake committee enforces the management of surface water on tributary river areas and Lake Jipe. Village government members, as well as non-members, can be members of this committee.

Although village water committees focus on the allocation of water for domestic consumption, they also enforce the protection of the water sources from which people draw water for this purpose. Since there are no tap water services, people sometimes use water sources situated far from their homes for domestic consumption because the water in furrows closer to homes is contaminated with cow dung and mud due to agricultural activities. The committee monitors water sources used for consumptive purposes and ensures that they are not contaminated by human activities, including farming. Sanctions are stipulated for non-compliance.

The village water committees are active in enforcing sustainable water management at Lake Jipe. They can even sanction village leaders if they do not comply. For example, the wives of the Village Chairperson and the Village Executive Officer were caught cultivating within the prohibited distance from a water source. There is a rule that when one violates regulations, he/she is charged a fine of TZS 10,000. When the two women were caught, they were required to pay the stipulated fine, but they declined. The case was taken to the village leaders, i.e. the Village Executive Officer and the Village Chairperson. Since the violators were their wives, the two leaders ignored the case. The chairperson of the village water committee took the case to the Ward Executive Officer. The two village leaders were called, reprimanded and paid the fines. This example indicates that when community actors are supported, they can have the power to participate in the management of natural resources together with governmental actors.

The lake committee enforces conservation of Lake Jipe from siltation from human activities. Among their responsibilities is to enforce excavation of soil harvesting structures at specified distances from the shore of the lack, so that soil eroded from agricultural practises is harvested from both the lowland and upland areas near Lake Jipe. As mentioned earlier the committee was mobilised by MIFIPRO. Nevertheless, village leaders govern the committee, and the community is involved in the excavation of soil. The long-run maintenance of the ditches is uncertain, however. For example, per metre excavated, farmers are paid TZS 1000. The participation of local people, therefore, may be motivated by the money they receive from this work. While this arrangement currently offers incentives for farmers to participate, in the future farmers may stop participating when they are no longer paid to continue the practise. The collaborative management of natural resources, therefore, must consider trade-offs between participation, incentives, and long-term ownership of initiatives, especially when the activity is undertaken in the style of a project, as is the case for MIFIPRO.

5.5.5 Customary institutions

Customary land management arrangements in the area of Lake Jipe are those institutions that operate within specified social-ecological boundaries and are legitimate to specific social actors. They are based on the traditional belief that if one does not comply to certain resource use practises, there is a danger of being stricken by misfortune or disaster. On customary land, there are clan forests that are used for traditional functions, such as rituals and initiation ceremonies. It is forbidden to cultivate these areas due to the value attached to them by the clans. These beliefs may seem irrational, but they have been a method of conserving water and forests in the area. At Lake Jipe because of the high level of legitimacy the *Pare* people attach to these arrangements, one can find governmental forest that is encroached upon while a traditional one is left untouched. The meanings people attach to natural environment around them can be useful in conserving the environment.

Traditional management systems operate in a centralised fashion, whereby a clan elder monopolises the decision-making and others are required to comply. These systems are male dominated and discriminatory in nature. The clan elder is always a man, and women are never allowed to approach ritual forests, no matter their age. The rationale behind this prohibition is that they might be at menstruating time which would upset the gods. Although such institutions can be helpful in conserving natural resources, they are also discriminatory, and therefore not consistent with the values of participatory natural resource management.

5.5.6 Irrigation farming institution

The irrigation farming institution is an informal organisation of farmers that participate in irrigation farming downstream of Lake Jipe. This organisation evolved from the common challenges encountered while carrying out irrigation practises. In other words, experiential learning caused farmers to organise as a mechanism to counteract emerging constraints. In particular, this arrangement developed from difficulties arising in the use of water resources for irrigation downstream when the resource is negatively impacted upstream.

Downstream irrigation practises are conducted based on allocation of water among farmers using shared furrows. The users of each furrow form an informal group that has leaders who mediate water use issues within the group. The leaders are essential for overseeing water allocation schedules because it is not possible for all farmers to irrigate at the same time. Furrow groups are led by traditional rules arrangements, which are respected by users since they are a legacy, inherited from their ancestors. Violation of these regulations results in being sanctioned in accordance with their norms. The furrow groups established an informal committee called the furrow committee, which is an informal organisation of farmers to address inter-furrow groups concerns pertaining to furrow use and the allocation of water. These committees are composed of leaders from various furrow groups.

Due to geographically narrow valleys where irrigation practises can be conducted upstream, some farmers from this locality have acquired irrigation plots and are members in irrigation organisations. Their presence is helpful in monitoring water use upstream when the water flow downstream slows. They act as informers about water problems after spying on upstream farmers' water use. If the cause for declining water flow from upstream to downstream is predominantly the poor practises of upstream farmers, the irrigation groups downstream establish an informal collective organisation to negotiate the situation with the upstream water users. The informal organisation enables resource users to negotiate for the equitable distribution of water resources, and through this process, a co-management arrangement has resulted between the upstream and downstream farmers, village governments in both areas, and customary institutions. Water used for irrigation purposes becomes the unifying factor in the negotiation process.

Practises that threaten the socioeconomic benefits to some users result in mobilisation for negotiation between water users downstream and upstream. This process facilitates the enforcement of formal conservation institutions because governments on both sides are involved, especially when farmers experience negotiation deadlock. Although the objective is not the conservation of water resources specifically, downstream farmers become a bridge for monitoring and enforcing water conservation practises and involving various actors in the negotiation process, and thus for enforcing conservation of water upstream.

5.5.7 Influence of MIFIPRO in the organisation of irrigation farming institutions

Although informal institutions govern irrigation practises, the non-governmental actor, MIFIPRO, has influenced the farmers' organisation at Lake Jipe. Traditionally, irrigation practises have been organised on an individual basis, and some individuals from various clans, especially elders, oversee and negotiate the allocation of water. The intervention of MIFIPRO has led to a shift in the farmers' organisation. This organisational shift developed after irrigation infrastructures were improved, which forced several farmers served by the same irrigation infrastructure to collaborate and negotiate the management of infrastructure and sharing of water resources. Group leaders have been established within individual groups, and a network of farmers groups (*UWAMAKA*) has emerged to oversee water use and management issues between the farmers groups operating at Lake Jipe basin. Over time, informal institutional collaborations have developed as a mechanism of sharing water resources among farmers. Although MIFIPRO did not introduced regulations to govern water use in the area, its intervention through the development of irrigation infrastructures shifted irrigation administration from individually-based to group-based.

Section 5.5 has uncovered how various formal and informal, as well as governmental and community, institutions and actors participate in managing natural resources in agricultural production. Whereas some institutions directly enforce the sustainable use of specific natural resources, as can be understood from their titles, for other institutions, the enforcement of sustainable natural resource management is an implicit part of a process in which actors negotiate the allocation of natural resource services to achieve socioeconomic goals, as was highlighted with the irrigation farming institutions. The section has also shown that some institutions integrate the government and the community (e.g. the environmental management committees, and the ward land councils), while others involve community actors enforcing formal rules (e.g. village water committee). Other institutions are made up purely of governmental actors (e.g. improved land use) who oversee compliance with the rules. In the next section, I will analyse conflicts that emerge in the use of natural resources in agricultural production and the conflict resolution mechanisms.

5.6 Conflict and conflict resolution

Conflicts can harm participatory natural resource management. It is impossible to create strategies for minimising conflicts, and therefore for increasing the potential for attaining natural resource management objectives, without first understanding those conflicts. This section will analyse conflicts that emerge in the use of natural resources in agricultural production and the mechanisms, actors and institutions used in resolving them.

5.6.1 Conflicts in natural resource management and mediation practises

Conflicts inherent in the use of natural resources in agricultural production at Lake Jipe area are land conflicts, categorised as farm boundaries conflicts (28%), land/farm confiscation (36%), and water conflicts between upstream and downstream farmers (35.7%), and among downstream farmers (31%) (Figure 5.8). Land is a competitive resource at Lake Jipe due to the presence of many users from various localities upstream and downstream conducting various activities. In the upstream area, there is limited arable land for food production. In this area, much of the land has been converted to permanent crop production, and some of the remaining land in the valleys is engaged in irrigation agriculture.

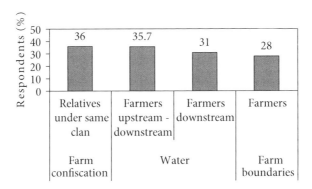

Figure 5.8. Conflicts types and involved actors in natural resource management at Lake Jipe (n=80).

5.6.2 Conflict resolution and mediating actors

Conflict resolution in the use of natural resources in agricultural production at Lake Jipe occurs through the direct involvement of governmental and community actors and institutions, and the indirect influence of a non-governmental agency. Direct resolution of conflicts involves the use of formal, informal, and customary institutions (Figure 5.9). The percentages in Figure 5.9 represent respondents' general ideas regarding the type of conflicts that occur along with natural resources use and management and the actors involved in resolving them. The respondents are not necessarily involved in these conflicts.

Customary institutions are the most commonly used in mediating land confiscation conflicts (59%), water use conflicts (55%), and farm boundaries conflicts (50%) at Lake Jipe. The government at the village and ward levels also participates in resolving natural resource use conflicts at Lake Jipe. The interviewees indicate that government participation in resolving these conflicts is relatively more common in land confiscation (41%), followed by water use (35%) and is low in conflicts related to farm boundaries (20%). Some institutions partake in resolving conflicts pertaining to one type of the resource due to their specialisation in issues related to this specific resource. Whereas some respondents (30%) indicated that Ward and Village Land Councils are involved in resolving conflicts related to farm boundaries encroachment, others (45%) responded that furrow committees mediate water use conflicts in the irrigation practises.

Generally, customary institutions are the institutions most commonly used in resolving conflicts pertaining to resource use at Lake Jipe. This use of customary institutions is due to the confinement of some natural resources issues into customary regimes. Due to the complexity and dynamic nature of customary management regimes and institutions in Tanzania (the country has more than 120 ethnic groups), the land law, as stipulated in the Village Land Act of 1999, has given power to customary land institutions to address land issues in the areas under their jurisdictions. The prevalence of customary institutions in resolving conflicts is also due to confidence people place in customary institutions, implying that they feel the elders give them a just result in the mediation process. Most problems that come to village governments go through customary actors

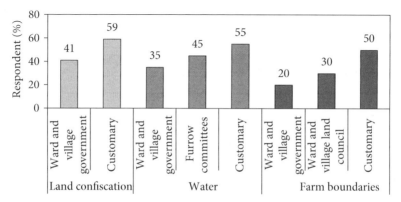

Figure 5.9. Institutions and actors involved in the mediation of resource use conflicts based on the experience of the respondents (n=80).

first and only advance to the government domains when they cannot be resolved at that level. However, there are areas in which the power of customary institutions is limited. Such is the case, for example, when conflicts emerge because of contradictory land allocation by village government. In such a case, the issue is under the direct jurisdiction of the government. In addition, even within customary arrangements, sometimes an accused person does not want to comply with the customary arrangements, and the mediators (elders) cannot force them to comply. Therefore, although the customary rules are applied under such circumstances, the government assists by exerting coercive mechanisms for the accused to comply.

Informal institutions also play a part in resolving natural resources use conflicts. Conflicts emerging in the use of water for irrigation activities at Lake Jipe are addressed using informal regulations governed by furrow committees. These address conflicts both within a single locality and between different localities (i.e. upstream and downstream areas). As mentioned above, conflicts occur within a locality when some farmers violate agreed-upon regulations, requiring them to observe irrigation schedules and arrangements. The violators are fined by elders or leaders of respective furrow organisations, but if they do not comply, they are reported to the village government or ward tribunal. These governmental bodies are able to enforce the regulations through the imposition of fines. For example, when unauthorised individuals use improved furrows, a fine of TZS 10,000 is imposed. A case can end with elders mediating between the conflicting parties, but if the accused do not comply, the case is reported to village and ward government to enforce compliance. Furrow committees also address conflicts between users of different localities, as well as within a single locality. These conflicts are resolved through negotiation with the elders of both localities. Similar to cases within a locality, if the farmers from one locality do not comply, the case is reported to the government at the village and ward level for coercive enforcement. The most common solution reached when water cannot flow to both localities at the same time, especially in severe dry seasons, is to schedule arrangements on a weekly basis.

Beyond the community level, the following government institutions partake in resolving conflicts: the Village Land Council, the ward tribunal and the district court. These institutions resolve conflicts among resource users, enforce and maintain peace and order among resource users, and protect the interests of user groups.

The Village Land Council, an institution appointed through the village assembly to administer land conflicts in the village, operates through an integration of formal land laws (as stipulated in Village Land Act of 1999) and informal rules governed by customary arrangements. The nature of conflicts determines which institutional mechanisms will be applied. For conflicts over land acquired through village governmental allocation, formal land rules are applied, whereas for land under customary land ownership, customary arrangements are applied. Since there are no written regulations for customary arrangements, clan elders are involved in the resolution processes while the land council may facilitate the process. Common conflicts of this type involve competition for the same land by descendants of same clan/family after their ancestors/parents have died, farm boundary conflicts, and encroachment on customary ritual areas for cultivation purposes.

The ward tribunal mediates resource use conflicts and enforces order and the conservation of natural resources in agricultural production. It also sanctions violations of the rules and regulations established among farmers and between farmers and other institutional actors in the management of natural resources in agricultural production. Sometimes farmers, in addition to violating environmental by-laws, threaten the security of enforcing actors, and this has to be dealt with by the ward tribunal when other institutions fail to contain the conflicts. If the conflicts occur on land under customary management, the tribunal resolves them in accordance with the customary rules. Sometimes, however, the ward tribunal may not have the power to resolve some issues. In this case, the district court is involved.

The district court addresses issues pertaining to land use, which cannot be resolved by the ward and village land instruments. For example, when customary land is encroached upon and the encroachers do not comply with the sanctions proposed under customary arrangements or with those applied at the ward level, the district court intervenes. However, although these conflicts are dealt with at the district governmental level, customary institutions still apply. Though the customary institutions are informal, there are provisions in the land law (Land Dispute Settlement Act of 2002) that provides for resolving land and other resource use conflicts under customary management systems.

The resolution of conflicts emerging in the management of natural resources in agricultural production is therefore the shared jurisdiction of the government and community institutions around Lake Jipe. Government intervention is necessary when community-level institutions, such as the farmers' irrigation institutions and other customary institutions, have failed to contain resource use conflicts. In most cases, however, conflicts are resolved at the community level involving traditional actors (the elders) and informal actors (irrigation committees).

Some conflicts are resolved using traditional rules because it would be difficult to resolve them using a formal rules framework. For resource use conflicts emerging under customary ownership systems, for example, even if the case is taken to the ward tribunal or district court, reference is made to customary rules in its resolution. Formal government regulations thus reinforce the customary rules. The latter are weak in that, for them to be enforced, they require the compliance of the accused person; otherwise, they do not have power to force individuals comply as the government does. While formal rules alone cannot resolve conflicts occurring in customary resource management systems, the customary systems do not have the power to enforce their own rules. This demonstrates that the government and the customary institutions are interdependent in resolving land use conflicts under customary tenure arrangements at Lake Jipe.

One actor who is neither a part of the government nor the community partakes indirectly in managing resource use conflicts at Lake Jipe. This actor is the non-governmental organization MIFIPRO. Conflicts in natural resource management within and between the upstream and downstream users may occur when the unsound use of natural resources at some locations cause the degradation of resources such as water and land, thereby creating a water shortage and pollution of water resources at other locations. Although the communities upstream and downstream of Lake Jipe recognise MIFIPRO as a capacity-building organisation, the role of MIFIPRO in improving farming practises in both locations may improve the use and availability of competitive resources (i.e. water) and reduce water degradation and pollution. These improved natural resource management methods indirectly reduce conflicts among natural resources users because when individual users in one area use natural resources appropriately, the resources are made available for other users in another area. Therefore, MIFIPRO plays an indirect but important role in mediating conflicts among resource users through improving their farming practises, which may in turn improve resource availability.

5.7 Co-management arrangement

In the co-management arrangement for the use of natural resources for agricultural production at Lake Jipe, there exist three types of institutions; namely, co-management institutions, informal institutions, and formal institutions.

There are three co-management institutions: the Village Land Council, the Ward Land Council, and the environmental committee. While the two land co-management institutions are narrow in their focus (i.e. they focus on land conflicts alone), the environmental co-management institution is wider in its focus because it covers multiple natural resources, including water and land, and works to enforce sustainable management rules in general.

The Village Land Council/co-management institution resolves conflicts occurring in the use and management of land at the village level. This co-management institution consists of customary and formal rules and informal actors (the villagers). While some actors are appointed through formal rules through the village assembly, others represent customary management systems in their specific clans as heads of these management systems. The village land co-management institution, therefore, mediates conflicts in land use by integrating formal and informal/customary rules.

In some cases, however, the Village Land Council is unable to resolve conflicts in land use. This is especially the case when violators or the conflicting parties do not comply with this institution. In this case, the issue is forwarded to another co-management institution, specifically, the Ward Land Council. The Ward Land Council involves formal and informal rules and integrates both governmental and customary actors. While the governmental actors are appointed through a formal Ward Development Committee (WDC) meeting, customary actors (elders) represent their clans. Both governmental and customary actors represent various villages in their wards. At this level therefore, there is an integration of formal and informal rules and governmental and community actors involved in resolving land use conflicts in the agricultural co-management arrangement.

The environmental co-management institution operates at the village level. This institution integrates governmental and community actors who are governed by formal rules. This institution governs and enforces the sustainable use of natural resources in the villages. The formulation of rules that govern the actors in the management process was influenced by the district government. However, following the rules formulation, the district government did not provide training to build the capacity of the actors in enforcing compliance with the rules within the community. Actors within this organisation do not have the ability to adequately enforce the sustainable management rules.

Another actor, the Mixing Farming Improvement Project (MFIPRO), fills this gap by training the actors who enforce compliance with environmental management rules. In this way, MIFIPRO builds the capacity of this co-management institution to make it function properly. The improved capacity of these actors to enforce sustainable management practises affects, among other practises, agricultural production. Therefore, MIFIPRO not only assists actors at the community level, but also builds the capacity of the environmental co-management institution. Without capacity building, the co-management institution would not be in a position to adequately govern sustainable resource use practises. This NGO is, therefore, not only a partner of the government in empowering community actors, but empowers the government as well because environmental committees integrate government and community actors. By empowering these co-management institutions, it consolidates vertical collaboration between the community and the government.

In the co-management arrangement in agricultural production at Lake Jipe, informal institutions also exist that govern the resource-use practises of community resources users and mediate their resource sharing and allocation relations. These institutions are resource-specific. They focus on water allocation, though water conservation may be indirectly facilitated in the process. Leaders in various users groups facilitate collaborations among informal institutional actors in a single location. Institutional collaborations also occur that involve actors implementing similar practises and linked by natural resources (water and land) but at different locations. These collaborations are facilitated by informal institutional actors who reside in one location but farm in another. Sometimes, negotiations among actors within the same locality and between actors of different localities under informal institutional arrangements do not yield desirable outcomes. Informal actors who assume leadership positions seek the assistance of governmental actors in both localities who have the power to enforce compliance with informal institutional mechanisms and decisions. In one specific case, such an action resulted in irrigation schedules between farmers upstream and downstream, especially for seasons when water was inadequate. Although the primary purpose of these negotiations is the allocation of water resources for irrigation activities, the constant monitoring of the water-use practises of the farmers upstream by the farmers downstream also results in the enforcement of sustainable water use practises upstream.

Another institutional collaboration of farming institutions in two geographically separated areas, and which focuses on water and land resources, occurs through the facilitation of MIFIPRO. MIFIPRO empowers community actors involved in the use of natural resources in agricultural production at the upstream and downstream localities. MIFIPRO improves farming practises by training the farmers to conserve the environment at Lake Jipe Since the sustainable management of Lake Jipe requires that farming practises should be sustainable both upstream and downstream, MIFIPRO bridges the two areas through improved farming practises. In this sense, MIFIPRO considers that sustainable management of Lake Jipe can be achieved only if a systemic approach that integrates the upstream and downstream localities is adopted. The NGO thus facilitates the resolution of environmental problems at Lake Jipe in terms of the two separate localities and their relationship, which make up different spatial levels of intervention. It bridges the two localities, which are isolated administratively and geographically, but connected by water and land. Through improving natural resource management practises, MIFIPRO indirectly resolves conflicts among users of natural resources, especially when natural resources availability is improved. However, although MIFIPRO provides training on sustainable management practises, successful implementation of these improved techniques and methods, even when the willingness is present, is a function of the capacity, in terms of resources (e.g. financial and labour), of farmers to sustainably practise them.

The present co-management arrangement in agricultural production at Lake Jipe indicates that three types of institutions - co-management institutions, formal institutions, and informal institutions - exist in this co-management arrangement. While some co-management institutions focus on one resource (the ward and village councils) and involve the use of both formal and informal rules to mediate one institutional dimension issue (conflict mediation in land use), others focus on multiple resources (environmental committee) and involve formal rules to enforce a different institutional dimension issue (environmental conservation). Some co-management institutions (Ward Land Council and environmental committee) are overseen by governmental and community actors, whereas others (Village Land Council) are overseen by the community actors alone. Informal institutions (irrigation farming institution) entail norms, customs, and traditions that govern resource use practises and relations and sanction non-compliance. Informal leaders appointed by users oversee these institutions.

The existence of resource management institutions confined to individual resources, the occurrence of resource degradation at various locations, and the existence of conservation institutions with a low capacity to enforce resource conservation has led to NGO intervention. Whereas governmental and community co-management institutions specialise in specific resources (i.e. some co-management institutions focus on mediating land use conflicts, while another co-management institutions focus on mediating and negotiating water use among the users), the NGO links the two natural resource management systems and builds the capacity of actors at the institutional level. In other words, the NGO builds the capacity of actors on issues related with land management and water management at different spatial levels, and intervenes at the institutional level by building the capacity of the co-management institution (village environmental committee). The NGO takes on different roles depending on the context. By linking arming practises in both locations and raising awareness of the environmental regulations among governmental and community actors, the NGO intervention strengthens horizontal community management and vertical government-community collaboration. However, MIFIPRO is attached to formal institutions, practises and mechanisms in its capacity-building activities with community and governmental co-management institutions. It is rarely attached to informal and customary institutions, mechanisms, and practises. This may imply that MIFIPRO places more trust in scientific knowledge than in traditional resource management knowledge, perhaps because it observes that in this area, traditional resource management systems do not seem to significantly control unsustainable resource management practises.

Although the present co-management arrangement establishes and strengthens linkages among resource users, and between users and the government, it does not reflect ethnic relationships between resource user groups or culturally-based interactions with natural resources. In practise, there is commonly more than one culturally differentiated user group in the same natural resource environment. These groups interact permanently or on a seasonal basis, depending on the geographic distribution and availability of the natural resource. At Lake Jipe, this situation exists in livestock keeping, whereby there are two categories of ethnic livestock keepers. So far, the present co-management arrangement in agricultural production has not reflected ethnic heterogeneity because it has not been understood to be a major issue in management conflicts. Similarly, although the co-management arrangement in fisheries acknowledges the existence of different user groups based on their localities (residents and non-residents), no formal or informal arrangements have been designed specifically to mediate the use of natural resources by different groups. The next chapter therefore will analyse the management of natural resources in livestock production, which involves two ethnic groups; the *Pare* and the *Maasai*. The chapter will discuss the ways in which these ethnic groups interact with each other and with the government in the management of natural resources in livestock production.

Chapter 6.
Management of natural resources in livestock production at Lake Jipe in Tanzania

6.1 Introduction

Raising livestock in the Jipe area involves a diversity of actors: the community and non-governmental and governmental actors. Three community groups exist that are geographically and spatially distributed but linked by livestock-related transactions: the *Pare* livestock keepers in the upstream areas, the *Pare* downstream livestock keepers, and the *Maasai* people in the downstream area. Thus, there are two livestock-related ethnic groups. The Pare ethnic groups in the downstream and upstream areas are related through kinship, and the livestock keepers in the upstream area entrust their livestock to the *Pare* people in the downstream region due to a lack of open grazing land in the upstream area. Therefore, while they live in the upstream area, their ethnic social network with the downstream *Pare* has enabled the upstream people to exploit rangeland resources in the downstream areas. Whereas livestock are managed downstream, nonetheless, decision-making regarding the livestock flows from upstream to downstream, while the entrusted provides information on the status of livestock to the owner upstream. Entrustment also exists between the affluent livestock keepers downstream and less affluent livestock keepers, as well as between non-livestock keepers downstream. Maasai livestock keepers are another livestock-raising group and unlike the *Pare,* they practice nomadic herding – that is, they move from one place to another based on the availability of pastures.

The government actors include ward and village governments and the district government. Specifically, the district actors include the District Agricultural and Livestock Department, the District Natural Resources Department and the District Commissioners' office. A non-government actor (MIFIPRO) also influences livestock and natural resource management practices at Jipe by partaking in some technical activities.

The relationships and interactions among the mentioned networks, institutions and actors in the use and management of natural resources in livestock production determine whether the management of natural resources in the Jipe area is sustainable or unsustainable. This chapter aims to analyse the interplay of various actors and institutions – governmental, non-governmental and community – in the management of natural resources for livestock production in the Jipe area. In particular, the study first looks at local institutional networks and the organisational practices of the livestock keepers at Jipe in exploiting natural resources. Second, it looks at strategies for improving the use and management of natural resources in livestock production. Third, it analyses how the government and the community actors interact in enforcing sustainable management institutions and practices in livestock production. Fourth, it considers how the two domains (the government and the community) interact in addressing disputes related to livestock and natural resource management. Fifth, the chapter identifies and evaluates the existing co-management arrangements.

As a start, an overview is provided of areas where the management of natural resources in livestock production is practiced, including problems and challenges as well as existing rights.

6.2 Grazing areas, challenges and existing rights

Table 6.1. Areas where livestock keepers residing in the downstream areas graze their livestock (n=80) (source: field survey, 2007).

Grazing area	Respondent (%)	Season	Reason
Communal land	100	During and after the rainy season	There is pasture at that time
The slopes of the mountains	100	During the dry season	There is no pasture for the livestock as part of the common land
The shore of Lake Jipe	100	During the dry season	There is no pasture as part of the common grazing land
Farm plots	43	After the crops are harvested	There is straw from maize and beans that is fed to livestock

Livestock grazing at Lake Jipe occurs on communal grazing land along the Lake Jipe shore, in the mountainous forest areas, and in farm plots after crop harvest (Table 6.1).

6.2.1 Communal grazing land

Communal grazing land is the land recognised by village governments as legitimate for herders to use to graze their livestock in their respective villages. It is the land allocated for livestock grazing. This land is owned by village government (a trustee) but is under the use and management of livestock keepers within specific villages. The livestock keepers hold the use rights to the land. Every livestock keeper in the village is entitled to access grazing land for grazing purposes.

In the Lake Jipe area, nevertheless, this tenure arrangement currently exists informally, and its boundaries are unclear. Because of this ambiguity, the common grazing land is, in practice, viewed as a land reserve that can be shifted to other uses when need arises. There are cases in some villages where this land has been given to other user groups: for example, in the Butu village, where the livestock keepers complain that the district government gave part of their grazing land to upland inhabitants who wanted to begin farming crops in the area. The livestock keepers resultantly express negative attitudes toward the district government, as the following response from an interviewee shows:

> *The government does not do us justice. It looks down upon the livestock keepers. We are also not adequately involved in decision-making. It is easier to bring in people from upstream areas to carry out farming practices and take us off our land on the basis that they are going to produce food crops. We don't have a solution because the government is powerful; otherwise, we would not allow people to simply come to our villages and take our land (Salehe Mfinanga, Butu Village, 13/6/07).*

The re-appropriation of communal grazing land is easy because of unclear grazing land boundaries. The above example of a district government's decision regarding grazing land in the village contradicts decentralisation policy because it is not clear who owns this land. Is the owner the village government or the district government, the latter of which made this decision? The ambiguous land tenure arrangements result in conflicts between the livestock keepers and other users, particularly farmers. Presently, however, the formalisation of the grazing land in Tanzania, including the Jipe area, is occurring, whereby village land is surveyed and boundaries are marked for various land uses to avoid such ambiguity. The district government land department engineers this process.

The Jipe area is part of a region frequently stricken by drought, especially during dry seasons. Then, the communal grazing land includes scarce pastoral resources. This has led to strategies on the part of livestock keepers for grazing their herds in other areas, as we see in the following sub-sections.

6.2.2 Grazing on the mountains and forest land

Grazing in mountainous and forest areas is one of the ways that livestock keepers cope with a shortage of pastures in the communal grazing land during dry seasons. Forest resources in the mountains are under the district forest authority. The use of these areas, however, is a violation of forest by-laws. Although this land is under government control, the land appears like open access management regimes because encroachment occurs for both livestock grazing and farming.[3] The government (district forest authority) has failed to protect these areas, likely because of limited staff and inadequate financial resources. Along with grazing, livestock keepers set fires to eradicate tsetse flies and promote the regeneration of fresh pasture. Fires, nonetheless, affect forest and water resources by degrading the vegetation cover and drying water sources. Besides, the movement of large herds of livestock in these areas also makes land vulnerable to soil erosion, and the eroded soil pollutes water sources. When it rains, the eroded soil is washed downstream to the Lake Jipe wetlands, contributing to siltation of the lake.

The failure of the district forest authority to protect the forest perhaps calls for the cooperative management of the forest resources in this area. The village community can be involved in this suggested regime. However, this would probably require justification as to how the villagers would benefit from participation in the regime; otherwise, the regime could hardly be successful.

---

[3] Researcher's personal field observation.

6.2.3 Grazing along the Lake Jipe shore

The presence of pastures on the shores of Lake Jipe attracts livestock keepers to graze their herds, especially during dry seasons, when pastures in the common grazing land are limited. The institutions, nonetheless, prohibit livestock grazing in these areas. The non-compliance with by-laws is attributed to the lack of adequate pastures due to drought spells. Livestock grazing practices on the shore of the lake, however, are associated with practices detrimental to the environment and resources. As in the mountainous areas, livestock grazing practices along the lakeshore are associated with fires used to encourage the regeneration of new pastures and eradicate the presence of tsetse flies. Likewise, the soil is eroded and compacted, and cow dung is deposited into the lake. This might be contributing to the flourishing of waterweeds due to an increase in nutrients deposited into the lake.

Although there are by-laws for environmental conservation in the villages, in practice, operationalisation is less effective in the Jipe area. This is probably due to the government's failure to provide livestock infrastructures in the area. The District Agricultural and Livestock Department had promised to construct livestock watering points and dipping facilities, as well as to demarcate and develop the pastoral land in the area. From the perspective of the livestock keepers, there are intense expectations for the government regarding the installation of livestock infrastructure so that unhealthy practices of grazing and watering livestock at Lake Jipe can stop. Not only do the villagers hold such expectations, but even the village leaders believe that the implementation of the environmental by-laws will be possible after livestock infrastructures are in place.[4] The village leaders sympathise with the lack of adequate infrastructures and resources for the livestock in their areas, and they excuse themselves for their ineffectiveness in enforcing environmental by-laws.

Other livestock keepers' practices, however, have some inherent merits. Livestock keepers burn waterweeds to open areas for livestock watering at some spots along the lake. This small-scale strategy, nonetheless, offers only a temporary solution and actually promotes the vigorous regeneration of the waterweeds. It appears that fire stimulates the growth of waterweeds, though further research will be necessary to affirm this. Nonetheless, based on the present case, we can learn that when socio-economic interests are interwoven with ecological problems – in other words, when a socio-economic problem is simultaneously an ecological problem – an attempt to address the former can at the same time assist in addressing the latter, at least on some affordable spatial and temporal levels. The livestock keepers at Jipe address environmental problems indirectly, though their primary intention is to obtain socio-economic (livelihood) benefits.

Regardless of the small-scale nature of livestock keepers' initiatives and the temporary nature of the solution, this case has some useful implications. It teaches government and decisions-makers outside the community that to achieve sustainable resource management, strategies for solving ecological problems should be used that rationally address and integrate inherent livelihood interests and motivate the community to act.

---

[4] Interview with the member of the village environmental committee.

6.2.4 Grazing on crop fields

Post-harvest residues are short-term feeds for livestock. This is possible after harvesting agricultural crops, usually around August and September. Though this offers a temporary alternative to animal feeds, it also means removing organic matter that can decompose to improve land fertility. By the same token, livestock trampling in agricultural fields can result in the deterioration of land, making it susceptible to wind and water erosion. On the other hand, this practice can result in gender-based conflicts between women farmers and livestock keepers. Interviewed women farmers complained that some male livestock keepers graze on their fields without their consent. When required to stop, the livestock keepers dismiss the farmers' requests and look down upon them, telling them, 'You are just women; you had better keep quiet'.

6.2.5 Land rights

Looking at the four natural resources areas where livestock grazing at Jipe occurs, as discussed above, one can see that there are three types of land property rights involved: integrated rights, state rights, and private rights.

Integrated rights are the rights to communal grazing land wherein both the government and the livestock keepers have some power. The government owns the grazing land, whereas the livestock keepers monitor its status and management because they are the daily users. However, as has been introduced, this property rights arrangement is currently insecure on the side of the livestock keepers because the government can alienate part of it and give it to other users, such as crop farmers. Furthermore, although the village government claims ownership, sometimes decisions are imposed on the village at the district level. Although on the side of the livestock farmers, dissatisfaction prevails over such decisions, especially when they have to do with grazing land, the power to resist is low.

State property rights exist in areas where livestock keepers are not allowed to exploit the land. Such areas are the forest area under the district department of natural resources. However, in practice, the district department is not able to protect this property rights arrangement, and livestock keepers continue to exploit the prohibited areas.

Private land right arrangements occur with individual farm plots. In this sort of arrangement, farm owners graze their livestock after crop harvesting season. An individual agro-pastoralist grazes his livestock on the farm, or a cooperative of livestock keepers pools its livestock and grazes it together on its farms based on informal cooperation.

So far, I have analysed natural resource use in livestock production in this section. In the next section, I will uncover and analyse organisational arrangements and pastoral networks existing in the use of natural resources in livestock production.

6.3 Local pastoral networks and organisational practices

Livestock grazed in the Lake Jipe area are the livestock kept on the downstream area and those entrusted by the upstream livestock keepers to the downstream inhabitants. This section, therefore, will consider pastoral networks at Jipe and their rationale, as well as livestock organisational practices in the downstream area where free-range grazing is practised. Understanding these networks and organisational arrangements is critical for gaining knowledge regarding who is involved in and what methods are used for the sustainable (or unsustainable) management of natural resources in livestock management.

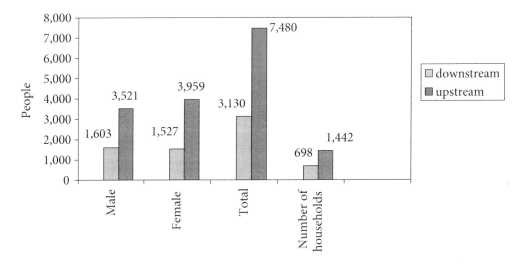

Figure 6.1. Population of upstream and downstream wards around Lake Jipe (source: National Census, 2002).

Table 6.2. Livestock keeping methods upstream of Lake Jipe (source: field survey, 2006; NA = not applicable; * = for chickens only).

Livestock grazing method	Respondents (%), n=110	
	Upstream	Downstream
Tethering	43.8	NA
Zero grazing	73.4	NA
Free-range grazing	100*	100

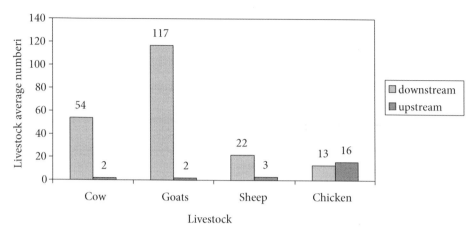

Figure 6.2. Average livestock per household between upstream and downstream of Lake Jipe (source: this field survey; n=110 households).

6.3.1 Local pastoral networks

The livestock existing in the Lake Jipe area is the combination of livestock moved from upstream and owned by downstream inhabitants. A large human population upstream (Figure 6.1) has resulted in land's being converted for the cultivation of permanent food (banana) and cash crops (coffee, cardamom).

This trend has resulted in a lack of land for practicing free-range grazing, and thus, indigenous livestock initially freely grazed the upstream has either had to be sold or had to be moved downstream. In turn, the upstream inhabitants have instituted improved zero-grazing, tethering livestock management systems (Table 6.2) and had to keep few improved dairy cows and goats and few sheep (Figure 6.2).

The relocated livestock from upstream are entrusted to inhabitants of the downstream areas where free-range livestock management practices are possible. Whereas some livestock keepers moved downstream and established settlements there, others remained upstream but entrusted their livestock to residents of the downstream area. In an entrustment, an agreement is made based on which when livestock increase, an entrusted person is paid in kind, usually with a kid or a calf at the third parturition per unit of livestock. Not only does entrustment exist between upstream and downstream people, but this arrangement also exists among inhabitants of the downstream area.

Entrustment involves livestock owners and people with few or no livestock. They are informal agreements between two parties; one (the entrusted party) invests his labour and time to keep and care for another person's livestock, and the other (the owner) pays in kind as the number of livestock increases. Again, there is an added advantage for the entrusted party. He enjoys other benefits such as milk and manure. There are opportunity costs to this arrangement, however. The entrusted person (always a man) has limited time to participate in other livelihood activities, such as agricultural activities, and thus often become a burden to a woman (his wife).

There are some prerequisites for entrustment arrangements. The entrustee must be a *Pare* inhabitant of the downstream area with permanent residence in the area, and the owner must trust him. The livestock owner monitors the reproductive status and increase of the livestock every year by paying visits to the entrustee and through information received from neighbours. Usually those with large livestock herds (more than 200) entrust their livestock to several individuals. Entrustment arrangements mainly involve rich livestock keepers, poor livestock keepers and those with no livestock at all. This classification of livestock keepers is based on the villagers' categorisation. Besides rich and poor livestock keepers, there are also average livestock keepers.

The rich livestock keepers are people with a large number of livestock, 100 to 200 or more (or 300-400 in the case of goats). Other people keep their livestock, and the owners pay them in kind with a kid or calf, based on the increase of the livestock. Characteristically, the rich livestock keepers have a prestigious reputation. They have formed networks with village and ward leaders that help the leaders in addressing urgent situations. For example, when one district leader visited the Jipe village (during my field research), the village entertainment fund was empty. Village leaders approached a rich livestock keeper who gave them a goat, which was slaughtered for the lunch for the district leader and his delegates. Rich livestock keepers also form livelihood networks with poor people (who do not have livestock) in the village to take care of their livestock and then be paid in kind when the number of livestock increases.

The prestige of rich livestock keepers, however, can have an impact on the natural environment. These owners inherently enjoying seeing their livestock increase without selling them. Extension officers' advice regarding reductions in the number of livestock seems irrational to them.[5] They can only sell their livestock during emergencies, such as famine, to buy food. To them, therefore, livestock is a security or a safety net. It is not easy for them, under normal circumstances, to sell or slaughter their livestock.

The second group of livestock keepers, based on the villagers' classification, are average livestock owners. The term average (according to the Jipe villagers' classification) refers to livestock keepers with livestock numbering from 20-50 cows or from 50-100 goats. This is actually the major group at Jipe.[6] The management of livestock is in most cases done through by hiring a caretaker who is paid TZS 8,000 to TZS 25,000 on a monthly basis. The third group consists of poor livestock keepers. The definition of 'poor' in the context of the livestock keepers at Lake Jipe is that a person owns less than 10 cows and less than 20 goats. Those with large herds of livestock (rich livestock keepers) may still entrust their livestock to individuals in this category based on a mutual agreement that those individuals will be paid in kind when the number of livestock increases.

While entrustment constitutes a livelihood network among the involved parties, it is also a strategy for distributing risk, justifying the use of different communal lands in different villages, and relieving communal land in a particular area from over-exploitation. Here, distributing risk means lowering the probability of the whole herd's being infected or killed by potential disease epidemics. The use of different communal lands in different villages is an advantage that the owner

---

[5] Village and ward extension officers.

[6] The village environmental committee member, Mr. Hatibu.

of the livestock can enjoy through the entrustment arrangement. In essence, every villager, in any particular village, is entitled to use communal grazing land in that village. The entrusted livestock, thus, can have access to various communal lands that can act as a buffer during critical dry seasons. While land in one village may experience a pasture shortage, land in another village may be in a relatively better state. Significantly, the entrustment arrangement can be environmentally friendly. The distribution of livestock to individuals in different villages reduces the large concentration of livestock in one area. Thus, the entrustment arrangement, besides being socio-economically important, can also contribute to the sustainable utilisation of pastoral and land resources. This may only be the case, however, when the livestock are entrusted to individuals residing in different villages. In contrast, if the entrusted individuals reside in the same village, there is the potential for a large concentration of livestock in the same village. This concentration may cause land degradation in a specific common grazing area.

Because the livestock kept downstream are from both upstream and downstream, any programme for the sustainable management of livestock production (e.g. promoting the reduction of the number of livestock in relation to the available land and pastoral resources) at Jipe needs to reflect the relationship between the livestock keepers upstream and downstream. While the livestock are kept and grazed downstream (entrustment), decision-making regarding the adoption and operationalisation of such a programme may be in the hands of livestock owners upstream or downstream. In addition, because the entrusted people downstream acquire livestock through entrustment, they may not even inform the owners of the sustainable livestock management programme when they perceive a threat to their livelihood.

6.3.2 Grazing organisational practices

Several grazing organisational practices exist in the Jipe area, namely individual organisation, cooperative organisation, caretaker hiring, and migratory practices.

Individual and cooperative grazing and hiring a caretaker

Individual organisation is when a caretaker (hired, entrusted, or an owner) takes care of the livestock herd of an individual owner. Commonly, for reasons of cost-effectiveness, this organisational structure is used with entrustment and self-grazing in most cases because hiring a person might be expensive for an individual person. Cooperative grazing is when a caretaker (hired, entrusted, or in a by-turns arrangement) takes care of livestock from two or more owners who pool their livestock together and pool financial resources to pay the caretaker. With this arrangement, besides financial payments, the caretaker is provided with food from the livestock owners. That is, the owners informally agree on a schedule on which they will provide food to the caretaker. Cooperative grazing organisation is cost-effective in that the owners share the cost, which makes this strategy cheaper than if each hired an independent caretaker. Hiring a caretaker is especially prevalent with households involved in agriculture and without children who can take care of livestock, either because the children are attending school or because they are older children and have moved to towns to look for jobs. Remuneration for the caretaker ranges from TZS 8,000 to TZS 25,000 per month based on the number of livestock he takes care of and the negotiation process between him and the owners involved.[7] The hired persons are not inhabitants of the Jipe area, but they mostly come from livestock-keeping agro-ecological areas, including the Singida, Dodoma, and Arusha regions.

Migratory livestock practices

The availability of open grazing land enables free-range grazing in the downstream areas. During some seasons, however, herders experience pasture scarcity. Among the strategies used to cope with this situation is migratory grazing. Migratory livestock grazing strategies, as one way to cope with a deficiency of pasture during dry spells, releases some areas that have been under utilisation at least for some time to encourage the restoration of the pasture. A temporal shift is commonly practiced among *Maasai* livestock keepers, though *Pare* livestock keepers sometimes practice the same. A temporal shift occurs to other areas in search of good pastures for livestock, or based on previous experience livestock keepers have with pasture availability in some areas at such moments. As the population has spread throughout the Jipe area, nonetheless, the movement has created diverse interactions among different natural resource users, such as farmers and other agro-pastoral livestock keepers. These movements have been a source of conflicts with settled farmers and livestock keepers. The emergence of these conflicts is due to competition for grazing resources (water and land) and sometimes to migrating livestock's grazing on the crops of settled farming communities. Along with these conflicts, at dry spells the livestock might be concentrated around small pastoral resources and water sources, which might result in the degradation of water and land resources as the carrying capacity of a given land is exceeded. The livestock keepers also have a tradition of setting fires in areas where livestock have passed to control tsetse flies and encourage the regeneration of fresh pasture. This further intensifies the degradation of water, natural vegetation, and land resources.

One of the ways the government has tried to intervene in shaping the Maasai's culture of moving from one area to another in search of pastures is through the enforcement of *Maasai* children's attendance at school, as well as through compelling herders to introduce crops farming along with livestock keeping. Traditionally, the Maasai give their children informal cattle herding education starting early, when the children are at the age of four to five years (usually the age for elementary education in Tanzania). At this stage, elders train children in the job of herding cattle by assigning them to look after young lambs and calves. At the age of six to seven years (the beginning of primary education in Tanzania), the children look after old cattle and can accompany adult livestock keepers, whereas later they can take the livestock long distances on their own (Phillips & Bhavnagri, 2002). This has been a legacy perpetuated from one generation to another. Apart from children's not attending schools, it has not been possible for the *Maasai* to have access to livestock extension and other social services, as they stay in the wilderness and continue to migrate from one point to another with their cattle. The government's intervention is based on the rationale that, first, the Maasai movement disseminates cattle diseases, deprives their children of the right to access to education, and deprives vulnerable groups such as women and children of important social and health services. Additionally, they degrade environmental

------
[7] Applies for 2006/2007, the rate might have changed presently.

resources, such as forests and water resources in areas they move to, and fifth, the migration causes conflicts between the Maasai and other user groups such as settled livestock keepers and crop farmers (MLD, 2006). Since the government assumes that providing formal education to *Maasai* herders' children will make the Maasai settle, the Mwanga district livestock authority considers the training of livestock keepers to be important for the sustainable management of livestock and natural resources. This is thought to be useful for herders to collaborate with the government in the provision of livestock services in their communities, as the government is limited, in terms of human and financial resources, to service all livestock keepers adequately.

Although a strategy to settle the livestock keepers can be important for reliably accessing them, when the need arises, and settling conflicts among resource users, it needs to be considered with care. The *Maasai* keep large herds of livestock. Migratory tendencies in such a situation can relieve pressure on environmental resources in specific areas compared to large herds being concentrated in one area. Since the *Maasai* are not used to keeping livestock in one area, the government must strive to empower the herders to sustainably use natural resources in the settled areas so as to avoid the degradation of natural resources.

So far, I have analysed pastoral networks and organisational arrangements by livestock keepers in the use of natural resources in livestock production. Already some introduction has been given on how the government intervenes in finding solutions to situations that endanger the

Table 6.3. Part of district strategies for development of livestock production at Mwanga district for 2006 to 2009 (source: Mwanga district, 2007).

Strategy	Cost (TZS)	Source of funds			
		Community	District council	Central government	Donors
• Improvement of the livestock veterinary services	2,000,000	√		√	√
• Rehabilitation and increase of livestock dipping trough	75,000,000			√	√
• Restoration and increase of livestock charcos	270,000,000			√	√
• Training of livestock experts from the livestock communities	5,600,000	√	√	√	
• Allocation and development of grazing areas	38,475,000			√	√
• Consolidation and formation of livestock keepers groups	530,000		√	√	

management of sustainable natural resources. In the coming section, further analysis is given regarding government strategies to improve the use of natural resources in livestock production.

6.4 Strategies to improve use of natural resources in livestock production

Strategic planning is an important aspect in natural resource management. It may be used to improve socio-economic and environmental interests. Well-prepared plans may appropriately combine human and financial resources to improve livestock infrastructures and services and, in turn, alleviate negative impacts to the environment due to the lack of important livestock infrastructures and services. The Mwanga district livestock strategic plan realises that the provision of improved veterinary technical services, livestock infrastructure, the allocation and development of pastoral areas, and the formation and consolidation of livestock keepers groups (Table 6.3) are critical for development of livestock management.

6.4.1 Improvement of veterinary services

As for other areas in Tanzania, extension services have to be shared among providers and beneficiaries at Jipe. Capacity is low, though, as both sides are deficient in one way or another. While livestock keepers are not organised as a community to support one another in maintaining and procuring prerequisite facilities and inputs for the technical services to be offered, the government lacks sufficient extension staff. The livestock keepers are not unified, to the extent that some livestock facilities and arrangements, which were provided and initiated by the Mixed Farming Improvement Project (MIFIPRO), could not be sustained. Rich, average and poor livestock keepers would not agree on the amounts of money to donate for veterinary services because some of them were said to seek free rides by contributing less money despite having many livestock. The government, on the other side, is constrained by staff inadequacy. At the country level, there is great discrepancy between available and demanded extension workers (Figure 6.3).

At the local level, the Mwanga district is deficient of livestock extension workers by 61% (Figure 6.4), and the Jipe area does not have livestock extension officers at the ward and village

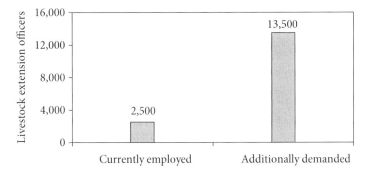

Figure 6.3. Comparison of currently available and demanded extension workers in agricultural and livestock sector in Tanzania (source: MLD, 2008).

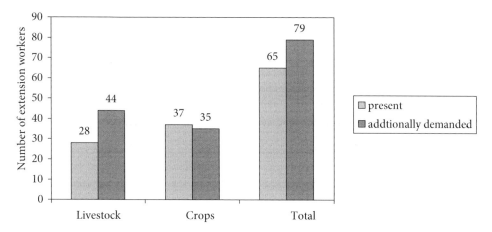

Figure 6.4. Extension workers present and additionally demanded at Mwanga district (source: Mwanga District, 2007).

Table 6.4. Extension workers at the Jipe ward (source: Mwanga district office, 2007; √ = available; x = not available).

Ward/villages	Agricultural extension officer	Livestock extension officer
1) Jipe Ward	√	X
2) Villages		
K/Simba	√	X
Kwanyange	√	X
Kivisini	√	X
Butu	X	X
Jipe	X	X

levels (Table 6.4). Livestock jurisdictions are taken care of by agricultural extension officers, whose competencies in livestock aspects are limited.

6.4.2 Training of livestock experts from livestock communities

In response to inadequate livestock technical staff, the government has devised a strategy to empower the livestock community to assume the roles of livestock service provision in collaboration with a few government technical staff (Table 6.4). This is important because in some areas there are no livestock experts. To achieve this, the government plans to train individuals from livestock keeping communities. This requires collaboration between the central government (the Ministry of Livestock Development and Fisheries), the Mwanga District Council, and the livestock keepers themselves as well as training institutes. However, for some livestock-keeping communities, such as the *Maasai,* the community has no formal education instrument for training, which makes it difficult to train them. For such communities, children are taught to keep cattle even from the elementary level onwards, by elders. Whereas they have practical knowledge on cattle keeping, gained from their experience in livestock keeping, they are short of formal education, a prerequisite for livestock training at the livestock training institutes. As a result, the government has recently declared that it is compulsory for migratory livestock keeping communities, including the *Maasai* communities, to take their children to school (TEN, 2007).

6.4.3 Improvement and instalment of livestock infrastructure

As has been mentioned above, in the provision of extension services, the government is deficient in extension experts and is financially limited in some areas. The failure to implement some services, nonetheless, can be attributed to not adhering to established work plans at the district

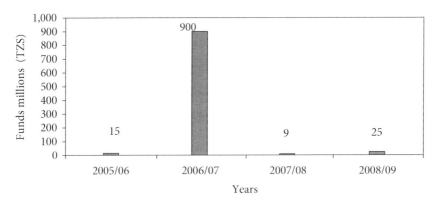

Figure 6.5. Money reserved for livestock infrastructures at Mwanga district from 2005/06 to 2008/09 (Source, Mwanga District 2008).

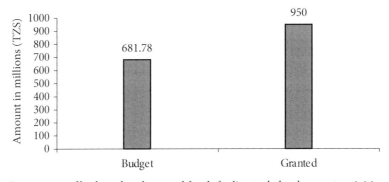

Figure. 6.6. Comparison of budgeted and granted funds for livestock development activities at Mwanga district for 2006-2009 (source: Mwanga District, 2008).

level. For example, an Agricultural Sector Development Programme (ASDP) through the District Agricultural and Livestock Development (DALD) invests in development of infrastructures among livestock communities (Figure 6.5). Specifically, the investment is for the rehabilitation of livestock watering dams, dips and charcos or the establishment of these infrastructures where they do not exist. The investment capital is made available to the district by the central government's Ministry of Livestock Development and Fisheries. Although the programme started in the Mwanga district in 2005, to date no livestock infrastructural development has been done at Jipe ward. The main reason for this lack of development given by the Mwanga District Livestock Officer is the scarcity of funds, which reduces the chance of serving many wards or villages at the same time. However, this excuse is somewhat controversial.

The comparison of budgeted funds to granted amounts shows that the granted amount was 39% higher than the budgeted amount (Figure 6.6). Although Mwanga district had applied for TZS 681.78 Million (for 2005/06 – June 2009) for livestock development activities, it was granted TZS 950 Million. It is likely that the established plans were not implemented. On the other side, livestock keepers continue to wait for the materialisation of a promise. They had been informed that the government would install livestock infrastructures (watering ponds and charcos) at Jipe. This fact was mentioned in interviews with livestock keepers and by the Makuyuni sub-village chairperson.

Whereas on the one hand, the inability to fulfil the promise creates mistrust of the government by the livestock keepers, on the other hand, the latter can use this weakness as an excuse to continue to exploit vulnerable areas for livestock practices. In participatory natural resource management, therefore, it requires the commitment to implement stipulated plans, not making promises that cannot be fulfilled, and when promises are made, there should be a dedication to fulfil them. When such promises are not fulfilled, it might hinder the other side, in the collaborative management initiatives, to implement its jurisdiction.

6.4.4 Formation and consolidation of livestock groups

Capacity building may need the formation or strengthening of local organisations among livestock farmers for drawing on and mobilising skills, especially when such organisations do not exist or are weak. These organisations become early adopters of innovations, and, as a cooperative, farmers can help one another in comprehending and disseminating skills among themselves and to other livestock keepers within and outside their area. The organisations can be especially important where there is poor community organisation for promoting and sustaining natural resource management technologies from the government or non-governmental organisations. At Jipe, the district agricultural and livestock department promotes the establishment of these groups. After the formation, they are imparted with relevant skills through training seminars and excursions to other areas where such activities are carried out. This is followed by the provision of property capital (e.g. goats) or financial capital for the groups to initiate participatory micro-projects and apply the skills and experience they have gained. A goat group already exist at Jipe. Although goats have been, and are still being, kept in the area, the government's intervention improves locally available goats through crossbreeding with improved billy goats. The district driven farmers' organisation at Jipe, however, puts much concentration on socio-economic benefits while the environmental concern is forgotten. The establishment of the livestock keepers group has not considered how the environment will be impacted along with the improvement of the indigenous goats. In most cases, the knowledge that the livestock keepers have is on keeping the goats and sharing of billy goats among themselves. Additionally, there is an overlooking and/or low involvement of technical people in the implementation processes of these projects. For example, the Participatory Agricultural Development and Empowerment Project (PADEP) at the District may advance directly to livestock keepers and implement some activities without going through technical and administrative levels at the village.

6.4.5 Improvement of the livestock welfare

Livestock welfare is important for the enhancement of the production and productivity of the livestock. Among important aspects for consideration in improving livestock welfare are protection against diseases and the improvement of the livestock pastures. The majority of respondents at Jipe (80%) highlighted that diseases affect the welfare of their livestock (Table 6.5).

The economically important diseases in the area include East Cost Fever (ECF), foot and mouth disease, eye disease, and fowl typhoid. Ticks are common vectors of the livestock diseases (53%). Diseases are controlled using both traditional and modern veterinary medicines. Medicinal plants such as *Azadirachta indica* (neem tree) are used in a traditional way. For the modern ways, the government through technical personnel imparts skills and offers technical services to the livestock keepers for the proper management of the livestock so as to improve livestock quality and health. For this method, livestock farmers have to cooperate with the government in improving livestock services. Whereas the livestock keepers are offered technical and veterinary advice regarding livestock, they have to purchase livestock medicine. The government is also responsible for offering extension infrastructures such as dip services, watering infrastructures, and charcos. This is an arrangement under the currently implemented Agricultural Sector Development Programme (ASDP) in Tanzania.

In addition to the agricultural activities outlined in Chapter five, MIFIPRO is also active in improving livestock welfare along with the government. It initially operated as a project but is now a Trust Fund. During the project time (1984-1994), MIFIPRO established a ward station for livestock services. It offered services including dipping, vaccination and treatment, and it established the ward livestock station. However, this station is presently not operational. After the end of the MIFIPRO project, village governments and livestock keepers within the ward were

Table 6.5. Problems associated with livestock keeping (source: field survey, 2007).

Problem	Household (%); n=80
Livestock diseases	80
Tick and tsetse flies	53
Wildlife attack	49

required to establish an organisation and prepare mechanisms for sharing veterinary medicines and operating costs for services for the livestock inputs and services to be locally available, as was the case during the project's tenure. Though the arrangement was established, it could no longer thrive.

Government and the livestock keepers accuse each other of failing to sustain the ward livestock station initiated by MIFIPRO. Whereas the village leaders claim that livestock keepers wanted the services free of charge, the livestock keepers assert that the ward and village leaders could not facilitate equitable organisation of cost-sharing arrangements among the livestock keepers. According to the village leaders, when the veterinary services were provided free of charge, the livestock keepers brought their livestock to the station. When they were required to contribute some money, they would not agree, and they stopped bringing their livestock for dipping, vaccination and treatment, and this led to the disoperation of the livestock station due to financial crisis. On the other side, the livestock keepers claim that their leaders contributed to the collapse of the livestock station. This group complains that the ward and village leaders did not rationally differentiate cost sharing among the livestock keepers. Livestock keepers with large herds of livestock were to pay the same amount as those with small number of livestock, and this resulted in conflicts between these groups of livestock keepers.

6.4.6 Improvement of grazing land

The improvement of pasture is critical for livestock keeping at Jipe area. A main problem faced during livestock keeping at Jipe surrounds the lack of adequate pasture for livestock, and this is the critical problem during dry seasons. There is even variability in pasture availability for consecutive dry seasons. For some dry seasons, the situation even leads to dying of the livestock. Such was the case in 2005. The Jipe area is recurrently stricken by drought situation. This is the main reason for livestock keepers' encroachment of fragile areas, such as forests and a shore of Lake Jipe. The Mwanga District Livestock Department understands this problem. Although among its objectives is to develop pastoral areas, it has not come out clearly how pasture improvement in practice will be done for the Jipe area. One of the ways the central government attempts to work to reduce deficiency of pasture in the rangelands is through the introduction of established pasture seeds in the rangelands (MLD, 2006). However, for this to be effective, rain is needed, so this might be effective during the wet season, but the same problem will persist during dry seasons. It appears that research on pastoral seeds has not been effective in developing pastoral seeds that withstand drought situations and can be prolific during dry seasons so as to reduce the movement of the livestock to other areas during these critical times.

The government, at least at the planning and advisory level, and to a limited extent at the practical level, collaborates with livestock keepers to improve the management of natural resources in livestock production. However, some issues exist between the government and the livestock keepers, which constrain, and may continue to constrain, the sustainable use of natural resources in livestock production. In the coming section, these issues will be examined.

6.5 Contrasting strategies between the government and livestock keepers

There is commonly an ecological concern, in areas where livestock are managed, on the relationship between the number of livestock and the amount of land or natural resources existing within a given physical environment (Fritz & Duncan, 1994; Scarnecchia, 1990). There is a discrepancy in the contemplation and way of thinking in this respect between the district livestock technical department and livestock keepers at Jipe. Whereas the district government relevant authorities uncover that the livestock population at Jipe is larger than a carrying capacity of the area - though they base their argument on theoretical grounds, as no study on livestock versus land area has been conducted in this area - the livestock keepers attach intricate socio-economic and socio-cultural reasons. These reasons motivate them to keep large herds of livestock.

The livestock experts at the district level claim that the livestock keepers do not take their advice of keeping herds in relation with the available natural resources. There seems, however, less understanding among the district technocrats of the rationale for the reluctance of the livestock keepers to follow the technical advice. I argue that unless socio-cultural and socio-economic environments nested in livestock keepers' decisions are understood clearly, strategies for improving the livestock and environmental nexus cannot be relevant and effective.

Among the *Pare* people, the livestock acts as a bank, as security assets, or as a cushion for times of hardship. During extreme droughts, for example, the livestock keepers sell some of their livestock in order to buy food. Likewise, they sell livestock when they encounter socio-economic problems, such as health complications of relatives, in order to address the situation. To the livestock keepers, first, the livestock act as shock absorbers, and as a cushion for these uncertainties. Second, at Jipe, as has previously been underscored, social livelihood networks exist between poor persons and rich livestock owners. Due to the lack of - or the limited number of - other options regarding livelihood, the poor who engage in these networks aim at procuring their own livestock along with increase in livestock number for owners. This is possible only when the number of the livestock increase, as their payments (a calf or a kid) are made on this basis. The villagers to whom the livestock are entrusted do not accept to reduce the livestock number, as this interferes with their livelihood objectives. As Maembe (2004) argues, the community at Jipe is not ready to embark on, or think about, environmental management while its members' livelihood needs are not met. This logic has to be properly understood before suggesting the reduction of livestock numbers. It is necessary to underscore the webs and chains of livelihood strategies and relationships within the livestock keepers' community so that feasible strategies can be suggested.

Aside from livelihood, socio-cultural function is another aspect of the necessity of the livestock among Jipe people. Livestock is used for mediating incompliance to some informal regulations. Likewise, it is used for an obligatory donation in a traditional arrangement where an individual during emergency is assisted by traditional social networks. The failure to comply with these traditional arrangements excludes an individual from the traditional social support networks. One such social network is *Ukunga*.

Ukunga is an informal institution under a traditional leader, who is usually an elderly person. One role of *Ukunga* is to facilitate and govern a social network of helping one another in case of emergencies. When the emergency occurs, the leader blows a whistle to alert people, after being informed about this situation by victims or associates. When the whistle is blown, every youth postpones what he is then engaged in and goes to the source of the whistle in order to receive a message over a prevailing incident. Traditionally, after the event, a household of the supported has to pay a goat, which will be consumed by all those who participated in resolving or attempting to resolve the emergency.

These networks have compulsive power to enforce actors to comply with their traditional arrangements. For example, one day during my field research, a group of youth, a sub-village chairperson, and the traditional leader came to the office of the Ward Executive Officer (WEO) with a young man. The young man was accused of a wish to commit suicide, by poisoning himself. He then disappeared into a forest to die there. The victim's mother yelled, which made the traditional leader blow his whistle to gather the youth to search for him. At this time, the victim was considered to have committed suicide. However, he was found alive. The victim's mother was required to pay a goat to those who participated in searching for him. However, the victim's mother could not pay the goat, saying that she did not have one. This violation of locally agreed institutions made the involved people present this case to the sub-village chairperson, who, nonetheless, could not resolve it. Therefore, the case was presented to the WEO, who listened to their claims. Since there were no provisions in the formal government institutions, the WEO advised them to go back and resolve the case using their traditional institutions. The traditional leader and his team declared that if the woman would not comply with their institutions, she would alienate herself from such support the next time she encountered such a problem.

The case above indicates that, the Jipe people attach not only livelihood and socio-economic implications to the livestock but also socio-cultural meanings. Social exclusion results from the failure to comply with culturally legitimised arrangements. The way that local people legitimatise such intricate socio-economic and socio-cultural issues implies that it is necessary to understand such issues before an outsider intervenes and suggests strategies for balancing livestock and pastoral land resource linkages in a particular environment. The livestock technocrats at the district level thus need to understand these relationships before they advise so that they can avoid providing simplistic solutions or advice with an uncertain likelihood of being adopted.

I have, up to this juncture, discussed livestock keepers' and government's strategies for the management of natural resources in livestock production. The discussion has unveiled areas where the two domains partake and contrast in the use and management of natural resources. In practice, the management of natural resources, besides comprising strategies and plans, is governed by rules and institutions, which direct how these strategies and plans should be implemented. The coming section turns to the enforcement of natural resource management.

6.6 Institutions governing use of natural resources in livestock production

The management of natural resources in livestock production occurs through formal and informal regulations, mechanisms and institutions. Some of these institutions are written down (formalised) under existing legal and institutional frameworks, whereas some emanate in the activities of the users as a way of defending, fighting for or securing their needs and interests. In the coming subsection, an analysis will be given on how various actors, the government and the community, participate in formulating the institutions (hereby meaning rules, by-laws, regulations, etc.), and how they participate in enforcing these institutions for sustainable natural resource management practices.

Some institutions emanate informally and others obligatorily emanate as a requirement from the existing institutional and legal frameworks of the country. The relevant institutions for the management of natural resources in livestock production at Jipe include environmental by-laws, forest by-laws, water regulations, and land formalisation regulations. Government and community interactions in the implementation of the environmental by-laws and water regulations have been discussed in section 5.5 encompassing all livelihood sectors, as the environmental regulations are multidisciplinary.

Forest regulations

Forest regulations fall into two categories. The first category of regulations is implemented by the District Natural Resources Department (DNRD) because they are applied for the protection of natural forest reserves that despite being within local areas (Jipe area) are under jurisdiction of the district. According to these institutions, it is not allowed to undertake grazing activities in these forest resources. Contrary to the currently promoted participatory natural resource management approach, though, the DNRD has not clearly stated how communities have to be involved, and the top-down approach continues to be maintained in managing these forests. Due to financial and technical staff limitations, nonetheless, the encroachment of these areas for grazing activities is not uncommon. Despite the fact that the district government enforces the protection of these areas, through awareness creation approaches such as the instalment of posters, the management is not working effectively.

Although approaches to joint forestry resource management have been in place in Tanzania since late 1990s (Kajembe, Nduwamungu, & Luoga, 2006), at Jipe the government continues to monopolise the management of the natural forest, which neighbours private lands. The forest department at the district seems to lack trust as to whether villagers can manage the forest, though at the same time, it has failed to enforce conservation of this resource using its top-down approach. On the other hand, monopolisation can be due to the fear of losing power and authority when these institutions would be shared with the community.

The second category of forest regulations applies to areas within village forests. These regulations are part of mainstream environmental by-laws and are currently planned to plant trees to combat degraded land areas in villages due to unsustainable farming practices. According to the village forest regulations, it is forbidden to graze on village forests. A violator of these regulations is fined TZS 50,000 (€ 28) or two years' imprisonment, according to the a member of environmental committee at Jipe. Due to the close monitoring of the forests at the village level, for the last three years, only two livestock keepers have violated this regulation; they had to pay the prescribed fine of TZS 50,000.

Land formalisation

One of the main reasons for land formalisation regulations at rural areas arises from the desire of the government to control for potential conflicts between farmers and livestock keepers, which have been commonly occurring in many parts of Tanzania (MLD, 2006). Currently, countrywide, there is an ongoing national programme of translocating the livestock keepers from one place to another so to control conflictual interactions between the livestock keepers and the farmers. The government has already moved some livestock keepers and livestock from regions, such as Morogoro (Kilosa district), Mbeya and other areas, to western Tanzania e.g. the Rukwa region and coastal regions of Lindi and Coast. This process has already caused a lot of havoc, both ecologically and socio-economically. Whereas the government, for example, states that the host areas would be improved with important social services, such as water, health, and education, much has not been implemented in practice. Again, socio-economically, the livestock keepers have incurred livestock deaths, loss of livestock, corruption and threats from enforcers of the translocation arrangement in the process (Mwarabu, 2007). Additionally, instead of alleviating conflicts, there is evidence of dissemination of the conflicts, whereby the livestock keepers in recipient areas compete with residents for natural resources exploitation (especially water and land) (MLD, 2006). Some farming communities already claim that the newcomers (livestock keepers) graze on their crops[8]. Instead of providing solutions for conflict experiences, as is common in the use of natural resources for livestock and agricultural production, the government-disorganised arrangement disseminates and escalates conflicts. It appears that there is poor planning in enforcing these regulations.

The Mwanga district participates in the countrywide campaign of demarcating livestock grazing areas from crop-farming areas. In the Kirya area, for example, the practice of demarcating livestock from agricultural (crops) land is underway. Contrary to Kirya area, nonetheless, for the Jipe area, the livestock keepers are not moved from one area to another. This is because a large percentage of livestock keepers at Jipe are also farmers (agro-pastoralists). Even the Maasai herders have started, though at a reduced scale, to incorporate farming activities in their livelihood strategies. For areas surrounding Lake Jipe, therefore, the formalisation of informally existing arrangement is practically what takes place. The formalisation of land use in this area aims at reducing conflicts among users and facilitating the sustainable use of natural resources.

Although the plan sounds good, it is yet disputable as to whether formalisation by itself is a solution to the dual problems of the unsustainable use of land resources and conflicts among users. Apart from the potential of formalisation processes to protect communal grazing land, the livestock keepers' encroachment of other areas, especially during the critical dry seasons, is likely to persist if pasture and water scarcity problems are not effectively addressed. This needs to be mediated among the government implementers on how the grazing land can be appropriately developed following the demarcation process so that the pastoral resources are made both sustainable and socio-ecologically feasible.

The budget constraint is an important factor that delays allocation and demarcation plans as well as the development of the grazing land. The local government and the community implement this programme by depending on funds from the central government, which is untimely allocated,

---

[8] http://www.ippmedia.com/ipp/radio1/2008/05/07/113911.html.

Table 6.6. Strategic plan for environmental conservation at Mwanga for July 2006 to June 2009 (source: Mwanga district, 2007).

Strategy	Activity	Cost (TZS). To be met by Mwanga District
Provide environmental education and enforce by laws against degradation and fire setting	• Train leaders at the village and district levels • Print and install awareness posts at the boundaries of forests • Prepare and disseminate brochures cautioning fire hazards to the forests • Sue the violators	3,850,000
To conserve water sources and enforce environmental by-laws	• Survey all water sources • Mark the boundaries • Plant trees • Sue violators of environmental by-laws	5,500,000
Total		**9,350,000**

and this has delayed implementation (Table 6.6). For example, whereas the idea of improved land use has trickled down to the villages, and already villages have started formulating by-laws for enforcing its operationalisation, up to 2008, when this research was being conducted, land surveys of the area had not started.

6.6.1 Enforcement of regulations/institutions

The government, the non-governmental actor (MIFIPRO), and the livestock keepers/community participate in enforcing the sustainable management of natural resources in livestock production in the Jipe area in one way or another. Formal and informal mechanisms interact in the enforcement process.

The informal organisation of livestock keepers as represented by a livestock keepers' committee aims at mainly defending their interests against other land users (especially protecting communal grazing land). Livestock keepers defend the communal grazing land from being encroached for farming activities. Through informing and convening meetings with village government and other actors (e.g. farmers), in case some interference occurs, they negotiate their interests and thus protect the grazing land.

Restricting the number of livestock from outside the village allowed to graze in communal grazing land during abundant pasture seasons (when other areas are short of pasture) is another informal way of regulating the use of communal grazing land. Although the government usually influences and intercedes on behalf of outside users, the decision regarding the allowable livestock number (from outside the village) into the communal grazing land is made by the livestock committee, based on the condition (as the livestock keepers assess) of their grazing land.

Shifting grazing during critical dry seasons is another informal arrangement that allows pasture regeneration for the communal grazing land. However, this institution creates land degradation vulnerability to other areas where the livestock are grazed, during these times (e.g. along the shore of Lake Jipe, on the mountainous areas, and long-distance shifts of the livestock).

Formal institutional mechanisms also enforce management practices of sustainable natural resources in livestock production at Jipe. These, as already introduced above, include environmental by-laws, forest by-laws, and land formalisation regulations. The government and local community enforce some of these instruments, whereas the government actor alone enforces others. The multidisciplinary committee (agriculture, livestock, and government), which integrates the community and government actors, enforces the environmental by-laws. On the other side, the water committee at Jipe comprises entrusted villagers alone, and its role is narrow, to enforce the conservation of water sources used for domestic consumption against encroachment and pollution by livestock. The forest by-laws (at the village level) enforce the conservation of forest resources. The forest by-laws are integrated within the mainstream village environmental by-laws. For forests under the district forest authority, nonetheless, the environmental by-laws do not apply, but regulations imposed by the district forest authority do apply and are enforced (though inadequately) by the District Natural Resources Office.

The District Natural Resources Department (DNRD), through its Forest Authority Section, made a plan to address the degradation of natural forests under its authority (Table 6.6). Its strategies include the provision of environmental education to leaders and communities, including livestock keepers. In addition, the demarcation of boundaries of water sources and forest resources is another strategy the district authority undertakes to combat forest degradation. The dissemination of brochures with precautions on fire hazards, and the impacts of the unsustainable use of the forest resources, as well as the attachment of awareness posters and signs at the boundaries of the forests, are other strategies to make people aware of these boundaries. If the villagers (livestock keepers inclusive) do not comply with these regulations, the plan states that punitive measures will apply, though it does not categorically state the type of punishments for incompliance. In the implementation of this plan, the DNRD understands that enforcement constraints exist, including the lack of adequate extension workers and the lack of adequate knowledge on how to make the by-laws.

Nonetheless, one can wonder whether the encroachment of the forest resources for livestock-keeping practices is because the community lacks awareness regarding the impacts of livestock keeping, setting fires on these resources, and the boundaries of the state forest resources vis-à-vis private property resources. Is this also because of their not comprehending the importance and role of the forests in providing environmental benefits and buffering hazards? Although up to the completion of this field research, (i.e. in 2008) the district government had not implemented a system of forest education for the community, the problem of encroachment of the forests for livestock grazing, and fire setting appears more attributed to the struggle to achieve social and economic benefits.

There is a discrepancy in the purposes and objectives between the district forest authority and the local livestock keepers. Whereas livestock keepers' objectives give first priority to securing livelihood benefits, the DNRD emphasises the implementation of the by-laws while externalising local people's interests. Since the DNRD cannot enforce its plans due to the deficiency, as earlier mentioned, of financial and technical human resources, the livestock keepers use this chance to obtain their livelihoods. The livestock keepers graze their livestock in or around the forest areas during dry seasons. This plainly indicates that at this time, the problem motivating the livestock keeper to graze the livestock in the forest is the inadequacy of pasture within the common grazing areas. I argue that the enforcement of the forest regulations by the DNRD has limited room for success unless reasons for setting fires and grazing on the forest areas are considered and alternative ways are discussed among the forest department and the local people. Such an approach can lead to the development of a feasible strategy and result in a decline in the encroachment of the forest. The integration of the government and livestock keepers' concerns can result in two benefits. First, the monitoring of the forest resources can be collaborative, which means that transaction and monitoring costs can be reduced. The second virtue is that the community can be provided with alternative and sustainable ways of meeting their livelihoods along with the conservation of the natural resources.

Along the enforcement of conservation institutions, socio-economic interests have to be integrated so as to increase the chance of successful conservation. This thinking has already entered the minds of district forest officers. The district has recently begun consideration of some incentive mechanisms to motivate villagers, including livestock keepers, to conserve the forests.

Enforcement through incentives

The Mwanga district's natural resources department appears to recognise that carrot motivation is imperative to incite the communities in the district including the Jipe area to participate in environmental conservation. Table 6.7 substantiates this argument. The district has initiated a contest for its rural and township areas to participate in tree nursery management and planting.

The tree-planting contest is relevant to the livestock keepers. The livestock keepers are users of the forest resources, although this is not legally allowed, on the argument that grazing in the forest degrades forest and water resources therein. Additionally, livestock grazing in the forest is

Table 6.7. Tree nursery management and planting contest at Mwanga district (source: Mwanga District Council, 2007).

Activity	Specific tasks	Funds required (TZS)	Sources of funds
Establish-ment of the tree nursery and tree planting	• Stakeholders seminar • Provision of quality tree seeds • Campaigning for tree planting in the villages • Assessment and awarding the winners TZS 500,000	• 200,000 for seminar • 6,000,000 for tree seeds and polythene bags • 500,000 for award	• The vice president environmental office • The Mwanga district council

associated with the setting of fires to kill livestock parasites, i.e. tsetse flies and ticks. The inclusion of the livestock keepers in this contest assumes that their awareness will be raised with regard to the importance of trees, and this might alleviate their negative practices within the forests.

The district underlies the following assumptions for initiating a tree nursery and planting contest: if tree planting is organised in the format of a contest, many actors will show up; there will be many tree nurseries and seedlings; the community will be motivated. So, in essence, according to the district' natural resource authorities, this implies that tree planting and conservation will be successful with the contest approach, whereby the winner is awarded TZS 500,000.

Whereas in the short-term this approach might be uneasy to dispute, in the long-term there are some uncertainties. For the participation of the people to be sustainable, the incentive arrangements have to be sustainable too, and not necessarily in financial terms. In the present case, if this incentive stops at some time, it is likely that the local people's attitude towards tree planting and nursery management will also shift. It will not be surprising to witness an impasse with regard to tree planting in the future when the present incentive system ends. Certainly, the mobilisation of the local people for conserving the environment is important, but perhaps it is also important to consider some types of sustainable incentives. In the context of the present financial incentive, the following questions are probably not easy to answer: will the government be providing finances of this kind on a sustainable basis? What is the implication of the incentive to perform environmental management in the long run, and what if the government offers an incentive only for a short period and then stops?

The awarding of tree-planting activity is an indication that the government is trying to use the carrot motivation approach, and on the other side, it implies that the stick approach is at least to some extent failing. However, there is no clear mention of the long-term planning of this arrangement. There is no clear analysis of the likely impacts of the incentive in environmental management, either. Thinking from a sustainability point of view, there can be uncertainty in the future as regards the participation of local people in the management of natural resources. The provision of short-term incentives might result in the abandonment of long-term strategies (Hellin & Schrader, 2003; IFAD 1996; Pretty & Shaxon, 1998). Under such situations, the empowerment of the community might not be emphasised, and this might affect the achievement and sustenance of long-term benefits after the end of the projects, which in this case are the awards (Giger, 1999; Hellin & Schrader, 2003). Positive responses that local people may show can be attributed to the urgency they feel to win the award. If this is the case, it is likely that they will cease to participate when the award arrangements stop. I argue that for people to participate continually in natural resource management, an incentive system should be well-constructed and devised based on considerations of sustainability. In other words, the environmental conservation objectives and incentive mechanisms should be mutually supportive. Otherwise, the response of the people can be temporarily positive just because of an existing incentive at a particular time, the lack of which can lead to poor or no participation later.

The source of funds for the contest and award is another aspect that requires consideration. For the operationalisation of the tree planting, the district natural resources office depends on financial aid from the Vice President's Office as one of the sponsors of the programme (contest). It is questionable as to whether this support is going to be sustained. Although incentive mechanisms can be an necessary tool for promoting participatory environmental management, it requires checks and balances between achieving short-term and long-term environmental benefits and sustainable livelihoods. A sustainable incentive system can build upon the livestock keepers' livelihoods. An example would be, along with tree planting, the integration of pasture development for ensuring that the livestock will get pasture for extended periods throughout the year and thus ensuring that fragile natural environments are protected from unsustainable exploitation.

In the Jipe area, a tree-planting contest has been held with the government at village levels, appointing some small groups to manage tree nurseries on behalf of the community. The village community will thus participate in tree planting exercises when the seedlings will be ready for planting. While the incentives can mobilise people to do certain tasks, especially after perceiving possible gains from the tasks, it might result not in understanding the necessity of environmental management but rather in the struggle to gain the incentives that have been offered to participants.

In the above section, I have analysed the institutions for enforcing sustainable natural resource management in livestock production in the Jipe area. There is an interaction of formal and informal institutions in the enforcement of sustainable management practices in livestock production at Jipe. Authoritative and carrot enforcement mechanisms exist. Authoritative enforcement requires that livestock keepers fulfil certain responsibilities, as required by the formal institutions, whereas the carrot approach is attractive when incentives are used to incite livestock keepers' willingness to participate. I have argued that, although socio-economic interests have to be considered along with environmental interests, care should be taken to check and balance the long-term and short-term environmental and socio-economic interests for the sustainability of these mechanisms.

Table 6.8. Conflicts between livestock keepers and other actors due to contrasting interests over natural resources.

Conflict type	Conflicting actors	Resolving/mediating actors
Competition for pastoral resources	Livestock keepers (*Pare* versus *Maasai*)	Village government
Crops destruction	Livestock keepers versus farmers	Elders Village and ward government District government
Land alienation	Livestock keepers versus district government	None
Wildlife attack	Wildlife versus livestock keepers	Villagers hunt and kill the wildlife

6.7 Dispute resolution

In the use of natural resources where users have contrasting or similar but competitive interests over these resources, conflicts are inevitable. At Lake Jipe, in the management of natural resources for livestock production, conflicts emerge in the competition for pastoral land for grazing livestock, for grazing versus food production, in land alienation, and between the livestock and wildlife (Table 6.8).

6.7.1 Competition for pastoral resources

Disputes emanate in the use of rangeland in the Jipe area. At critical seasons of pasture deficiency, the seasonal migration of livestock occurs along Lake Jipe between villages and areas. Although seasonal movement is not unique for *Maasai* livestock keepers, the *Pare* livestock keepers treat one another differently from the way they treat the *Maasai* herders. The *Pare* people apply reciprocity institutions among themselves for hosting one another during critical pasture seasons. In other words, they use the 'host me today then I will host you tomorrow' approach. The *Maasai* keepers conduct things differently. The *Maasai* livestock keepers move as far as the Jipe village from Toroha, Ndea Mgagau villages where they usually camp. Since other groups of users (the *Pare* farmers and livestock keepers) occupy areas that *Maasai* pastoralists move to, conflicts emerge between these two ethnic groups. Conflicts occur when the *Maasai* livestock either graze on food crops fields or browse on communal grazing land under other livestock keepers. Also, the *Pare* accuse the *Maasai* of stealing their cattle. The village, ward and district authorities address these conflicts. Conflict resolution arrangements start at lower government (village) levels, and when they fail, upper levels intervene. Because of the conflicts, some strategies endogenously develop to facilitate the harmonious use of natural resources for livestock production between the two ethnic user groups.

Legitimising the use of natural resources by a formerly excluded social group is one way of attaining the harmonious use of resources among different ethnic users. At the Jipe village, for example, the village government has informally developed a mechanism for legitimising the use of communal grazing land by the *Maasai* during seasonal movements from their camps to other areas. One such arrangement is the informal purchase of grazing rights depending on the number of days spent at the communal grazing land of another village. For example, for three months, the *Maasai* pay up to TZS 50,000 as a use rights fee. The payments are not charged based on the number of the livestock or on the area covered by the livestock but rather on the number of days spent at Jipe. Then, the government negotiates with *Pare* livestock keepers to allow the *Maasai* to graze on the communal grazing land with the argument that the *Maasai* contribute to development activities at the village.

The integrated government herders' property rights regime for the communal grazing land gives the government some power and influence over resident *Pare* livestock keepers. Although the livestock keepers use this land communally, they do not own it, as it is under the government (entrustee). The government can thus influence and recommend use by outsiders. The Maasai use the money to purchase grazing rights. Besides purchasing the grazing rights, when at Jipe village, they are also involved in fundraising programmes for participatory development activities (e.g. construction of schools). This was evident in an interview with a member of the village government, who asserted that:

> *The Maasai follow our regulations, they present their applications for being allowed to graze in our village land, and they pay for their application to the village council based on the number of months they are going to spend. They also contribute to development activities in the village. However, the livestock keepers determine how many livestock have to be allowed in the communal grazing land based on pastoral land condition. Sometimes when we ask them to host the Maasai, they respond that it is not possible because our grazing land is in poor condition. (Hatibu, Makuyuni Subvillage Chairperson, 27/6/07).*

Although the village government intercedes for the Maasai, nonetheless, after receiving a use rights fee, the *Pare* livestock keepers through their committee (informal organisation) decide on the number of livestock from the outside to graze on the common pastoral land, as evidenced in the sub-village leader's statement above.

Financial benefits attract the government to support the use of grazing land by outside users. The livestock keepers express dissatisfaction with the government's statement that the *Maasai* contribute to the development of the Jipe village. One of the livestock keepers in the area claimed that he does not know whether the use rights fee the *Maasai* pay goes to the development activities in the village or whether it is shared among village leaders. The lack of transparency on how the collected fee is used prompts queries among the *Pare* livestock keepers.

The Maasai and Pare, even though they are found around the Lake Jipe area, hardly intermix. The weak association between the two results from a lack of trust. A lack of trust can negatively affect a relationship between actors. This situation can occur when the characteristics and behaviour of one or several actors in the institutional framework create a sense of insecurity among or endanger the interests of other actors. The *Pare* livestock keepers at Jipe distrust the Maasai herders because of the latter's tendency to raid livestock. Traditionally, according to the Maasai's culture, cattle belong to them while other tribes who keep cattle were in the past given the cattle to keep them temporarily, so the Maasai have the right to take them back (Mbanile).[9,10] Because of this tradition, even in villages where both ethnic groups live, they do not mix, due to this uncertainty and fear among the *Pare*. The *Maasai* stay in their own camps, and the *Pare* stay in theirs. Likewise, for areas to which the *Maasai* seasonally relocate, especially during critical pasture deficit periods, they are socially excluded from coming and staying in the midst of the host *Pare* community because of this fear.

6.7.2 Destruction of other users' property

Conflicts are unavoidable when the resource use practices of one actor or several actors disrupt interests of another natural resource user. At Jipe, *Maasai* herders during their movements along

---

[9] http://fga.freac.fsu.edu/resources/academy/pdf/tribal_fact_sheets.pdf.
[10] http://www.wilsoncenter.org/topics/pubs/reportfromafrica12.pdf#page=16.

Lake Jipe area, in search of fresh pasture, graze on food crop fields of settled farming communities. This results in disputes between the *Maasai* herders and the farmers. In 2005, for example, disputes between the two groups were so intense that the ward and village governments could not resolve them, and consequently the *Pare* villagers confiscated the *Maasai* cattle. For such conflicts, when lower governmental levels have failed to reach a resolution, the district government intervenes. The district commissioners' office had to intervene and mediate between the two ethnic users of natural resources at Jipe, and the cattle were given back to the *Maasai*. In this category, conflicts also sometimes exist between the Pare livestock keepers and farmers. The browsing of food crops by the livestock is a reason for this. These conflicts hardly come to the government but are mostly resolved by elders in the *Pare* community. This indicates that intra-ethnic conflicts are within the reach of respective ethnic groups and that solutions to these problems can be obtained within ethnic boundaries by using instruments the community itself regards legitimate.

6.7.3 Land alienation

Scarcity and the ambiguous property rights of a competitive resource can result in resource use conflicts. At Jipe, as elaborated under subsection 6.5.1, the alienation of grazing land by the district government results in the hatred of the district government by the livestock keepers. This land is given to farmers from the upstream area because land is scarce in the area. Though the livestock keepers disapprove the alienation of grazing land, they do not have the power to prevent the government from doing so. The type of property rights for the grazing land limits the livestock keepers from resisting the government because the grazing land boundaries are ambiguous. Additionally, the government holds ownership of the grazing land, whereas the livestock keepers have use rights. This makes it difficult for livestock keepers to prevent the government from making the decision make grazing land exclusive for other uses. The district government thus uses its power to provide land to upstream farmers who do not have enough cultivation land on the upstream.

6.7.4 Invasion by the wildlife

In areas where the livestock-keeping community lives closer to national parks, conflicts between livestock and wildlife are difficult to avoid (Bangs & Shivik, 2001; Jackson & Wangchuk, 2001). At Jipe, this occurs during dry seasons, when water in the lake has declined. In the past, elders uncover, the Jipe area had many wildlife including lions, elephants, buffalo, and hippopotami. Elders highlight how in those days they fought lions as they grazed their livestock. The attack of livestock by lions was frequent. Some areas are named after this historic phenomenon. One of the villages in the Jipe area, for example, is called *Kambi ya simba*, literally meaning 'lion camp'. Settlement distribution in the area has displaced the wildlife. Since on the Kenyan side of the Tanzania-Kenya border there is the Tsavo National Park, the wildlife moved to this park. Nonetheless, the wildlife occasionally move into the Jipe area. The most common threat to the livestock keepers are lions. This problem is critical when water in the lake is low, as the lions move around, probably in search of drinking water. In 2005, when the lake encountered a severe drought, for example, the Jipe area experienced frequent livestock attacks by lions, especially at night. The livestock keepers encountered loss of their livestock. Mr. Mustafa, a sub-village leader, for example,

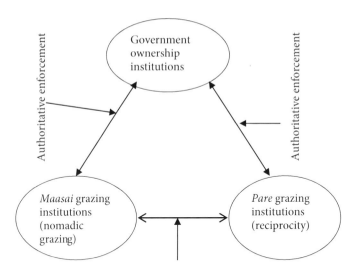

Determine, govern, mediate, and link mutual use
of pastoral resources among the two ethnic groups

Figure 6.7. Interactions among ethnic and governmental institutions in the use of pastoral land at Jipe.

lost seven cows in one night when the lions attacked his livestock enclosure. However, the existing solution is presently to hunt and kill the lions. There exists no formal collaborative arrangement in place to control this problem. Nonetheless, when it is critical the district's natural resources department assists in protecting local people and livestock against lion attacks by hunting the lions.

The experience of conflicts between the livestock community and wildlife has socio-economic and environmental connotations. Whereas people's lives and property are endangered by wildlife attacks, human interference in the environment contributes to this problem. The critical drying of Lake Jipe in 2005 was due to the abstraction of water at tributary rivers that discharge water into Lake Jipe for large-scale agricultural production. Such rivers include Kirurumo from the Pare Mountains, and Lumi, which enters Lake Jipe through Taveta, Kenya, although it originates from Mount Kilimanjaro on the Tanzanian side. The cost of degrading the environment can result in a sense of both environmental and socio-economic insecurity.

6.8 Co-management arrangement

The co-management arrangement in natural resource management for livestock production at Jipe is built on ethnicity (Figure 6.7). Relationships between the two ethnic groups, the *Pare* and the *Maasai*, are determined by spatial and temporal variations in the availability of pastures on the pastoral land. At seasons when pastures are adequate, the two ethnic herders become spatially distributed around the Lake Jipe area. While during scarce pastoral seasons some livestock actors move from one area to another in search of pasture and water, they meet and interact with other herders settled in some particular areas. Based on pastoral availability and scarcity, some ethnic-based local institutions have evolved among the users, whereby they recognise one another based on these ethnic differences, and even the treatment of individual livestock keepers at some areas follows these ethnic rails. Reciprocity institutions, for example, exist among *Pare* herders where, based on their homogenous ethnicity, herders in one village are allowed to graze in another village, on the grounds that the assisted today may assist others tomorrow or in the future. This acts as social capital and as a network for them. While reciprocal institutions exist among the *Pare*, they distrust the Maasai herders, based on the perception that the latter are cattle raiders. The mutual temporal use of grazing land for the two is thus facilitated through government-driven formal institutional interactions.

Formal property rights arrangements influence the interactions among social actors in the use of pastoral land at Jipe. Formal institutional arrangements for the pastoral land recognise this resource as state-owned, whereby, within administrative village boundaries, the government acts as a trustee of the pastoral land, whereas the herders have the right to use it. Decision making over the pastoral land, therefore, is a shared jurisdiction of the government and the livestock keepers. This institutional arrangement, existing between informal local institutions and governmental formal institutions, indicates that although the *Pare* herders have the right to use pastoral resources, the government holds the decision-making power that can influence the inclusion of other users, the *Maasai* in the present case, in the use of the resource.

On the other side, the *Maasai* herders have some influence that catalyzes the interactions. Whereas the government uses its power as the owner or trustee of the village pastoral land at Jipe to intercede for the *Maasai*, the latter have power in the form of financial resources, which make the government actively intercede on their behalf. As has been explained before, the *Maasai* are required to pay a fee (about TZS 16,500 per month) to be allowed to graze their livestock on the communal pastoral land at Jipe villages to *Pare* villagers and government administration. Thus, it can be argued that the governments of *Pare* villages benefit from this intermediary work and that this is at least one of the reasons motivating the government to intercede on behalf of the *Maasai* herders.

The new spatiotemporal co-management interactions among the *Pare*, *Maasai* and governmental institutional actors are thus facilitated by some factors that affect power relations among these actors. These factors are resources, rights and persuasive power (Figure 6.7). The government has rights as the owner of pastoral resources, and it uses this right to persuade the *Pare* livestock keepers to allow the *Maasai* herders to use the pastoral land. This persuasive power, however, is likely exercised because of financial resources the *Maasai* herders pay, which motivate the governmental actors to intercede on their behalf. Despite this resource-based motivation, however, the interactions among the herders' ethnic groups and the government indicate an important position that governmental actors can play in mediating interactions among multiple local resource users, especially when there are multiple ethnic groups involved. This co-management arrangement alleviates conflictive experiences between the two ethnic groups and enables the mutual use of pastoral land resources by the two ethnic groups, especially at the time when the physical environment has spatial scarcity of pastoral resources.

This case shows that in one institutional arrangement, with formal and informal institutions, there is a tension between culturally diverse informal institutions governing the use of natural resources by different user groups. Since the ethnic actors are in tension, in order to enable the collaborative use of natural resources, it requires a linking and mediating role of other kinds of institutions, that is, the formal (government-based) institutions. However, some ethnic users of the natural resources use (financial) resources to cultivate the mediating role of the government institutions by influencing an authoritative actor (government) to intervene and act in their favour.

Although co-management arrangements in the use of natural resources in livestock production has recognised the existence of ethnic groups and has attempted to harmonise the mutual use of land resources among them, the continued collaborative use of pastoral resources by these two ethnic groups depends on continued government intervention. However, the government intervenes because of the benefits that it gets from one of the ethnic users. This arouses uncertainty as to whether the government would favourably act for this actor in the event that the actor lacks the resources to influence the government. This indicates how difficult it might be to practice the collaborative and sustainable use and management of natural resources in light of social-cultural diversities. For the sustainable co-management of the pastoral resources, facilitation between the two ethnic resource users, rather than the current government intervention, is necessary. Although this may be hard to achieve, based on the fact that the government benefits from its intervention, sustainability inevitably requires users themselves to negotiate mechanisms for the mutual use of natural resources between the ethnic users' categories, while the government facilitates the negotiation process.

Up to this point, the three empirical chapters have been discussed independently of one another. In practice, the three co-management arrangements occur within the one socio-ecological environment, which is Lake Jipe. These co-management arrangements, institutions and actors interact and influence one other in various ways and as a whole influence the sustainable management of Jipe's socio-ecological system. In the concluding chapter the three co-management arrangements will be compared.

Chapter 7.
Conclusion: co-management of social-ecological systems

7.1 Introduction

This thesis has investigated the emergence of co-management arrangements for sustainable management of fisheries, agriculture and livestock at the Lake Jipe wetland in Tanzania. The results have highlighted the often overlooked dimension of intra and inter-sectoral linkages by making 'legible' (cf. Scott, 1998) the linkages within and between these sectors at different spatial and temporal scales, through a complex set of formal and informal institutions. As such, the thesis contributes to co-management theory by focusing attention on the dynamic interaction between what are termed co-management *arrangements, institutions* and *actors*. In doing the thesis has built on the existing work of, among others, Adger *et al.* (2005), Berkes (2000), Wilson *et al.* (2006), and Cash *et al.* (2006), in providing a clearer framework for analysing cross-scale and cross-sectoral networks of co-management that moves beyond community-government relations and highlight to complex governance arrangements in complex social-ecological systems.

The preceding chapters also provide an important empirical analysis through the case of Lake Jipe, which demonstrates the contemporary challenges of natural resource management in Tanzania, and East Africa more broadly. Like many natural resources in Tanzania, Lake Jipe wetland has been subject to centralized management. As outlined in Chapter 2, the limited capacity in human and financial resources of the government of Tanzania to sustain centralized top-down natural resources management regimes and the continued deterioration of natural resources at Jipe and other socio-ecological systems has led to the introduction of decentralisation and participatory natural resources management regimes since the mid 1990s (Kajembe *et al.*, 2006). This decentralisation process, of which co-management is one expression, has directly and indirectly (re)distributed responsibilities for management along government-community lines. However, as this thesis has demonstrated, co-management is comprised of multiple management arrangements within a complex social-ecological setting such as Lake Jipe. Making these arrangements legible, through the institutions and actors that both comprise and link them, provides a clearer understanding of the transition in Tanzanian natural resource management and a starting point for more informed policy.

In this final chapter a better understanding of co-management is addressed by answering the key research questions of the thesis. First, how do government and community institutions and actors interact in the management of natural resources in agriculture and livestock production and in fisheries management, and what is their relevance for sustainable management of the Lake Jipe social ecological system? Second, what co-management arrangements emerge from interactions involving governmental and community institutions and actors around livestock, agriculture and fisheries? And third, how are the three sectoral co-management arrangements related with respect to sustainable management of the Lake Jipe social-ecological system?

These questions are addressed in four stages. Section 7.2 summarizes the key characteristics and challenges faced by institutions and actors within the fisheries, agriculture and livestock co-management arrangements presented in the preceding chapters. Section 7.3 then reflects on the intra- and inter-sectoral linkages between what emerge as multiple cross-scale co-management arrangements and their relevance to natural resource management in Tanzania. Section 7.4 then broadens the findings of the thesis to discuss the theoretical implications of this thesis for co-management and the governance of complex social-ecological systems before the final section turns to the conclusions and recommendations.

7.2 Institutions and actors in three co-management arrangements

Multiple lines of conflict and resolution characterise the shifting relations between and among resource users and management actors around Lake Jipe. Co-management arrangements within each of the three sectors clearly highlight the source of conflict and the degree to which both new and existing institutions and actors mediate or enhance collaboration in the stewardship of shared natural resources, similar to what Singleton (2000), Sowman *et al.* (1998), Castro and Neilsen (2001) and Pomeroy *et al.* (2004) argue with reference to various natural resources in North America, South and Southeast Asia. The following builds on these studies by identifying the key characteristics and challenges of institutions and actors involved in intra-sectoral fisheries, livestock and agricultural co-management arrangements and the degree to which they dealt with resource use related conflict.

7.2.1 Fisheries co-management

Fisheries at Lake Jipe comprise diverse groups of fishers differentiated by geographic origin. Some are residents to the area whereas others, migratory fishers, come from southern Tanzania and other nearby regions. The inadequacy of the government and fishing communities to independently manage fisheries resources at a sustainable level around Lake Jipe led to the instigation of collaborative forms of fisheries management. In resolving the complex conflicts that ensue between these groups of fishers, formal and informal institutions have the potential to collaborate within and between independent geographic areas (cf. Jentoft, 1999). In particular, cross-scale interactions and collaborations have evolved entailing community and district governmental actors. Likewise cross-scale collaborations occur between informal community institutions and actors and formal authoritative governmental institutions.

A key example are the village fisheries committees around Lake Jipe. The various communities around the lake have been instructed by formal authorities to appoint fisheries committees to enforce fisheries laws. The fisheries committees, as co-management institutions, have encountered a series of challenges in enforcing fisheries regulations from both fishers and some government leaders who are responsible for unsustainable fishing practices and have allied with fishers using illicit practices. The poor leadership deprives the fisheries committee institutional support by siding with the illicit fishers in addition to degrading the fishery resources. These institutional failures, similar to the findings of Pomeroy *et al.* (2001), Neilsen *et al.* (2004) and Isaac *et al.* (1998), are exacerbated by the geographic distinction of those 'inside' and 'outside' of the committees jurisdiction around Jipe, and have led to threats of violence against the committee members.

Conflict between resident and non-resident fishers has also ensued, especially in relation to competition for prime, but limited fishing areas. Conflicts between resident and non-resident fishers are resolved with the aid of governmental authoritative institutions and actors because informal institutions – though legitimate in the eyes of the local community – do not have coercive power to enforce non-resident actors to comply. However, the informal institutions play an important, though sometimes small, role to channel information to government authorities on conflictive interactions between resident and non-resident fishers. Connections between formal and informal institutions facilitated by the fishery co-management arrangement have, as argued elsewhere by Lebel *et al.*, (2006), created new informational linkages which empower resource users. In addition, a degree of empowerment is developed whereby fishers are able to flexibly react (cf. Raik, 2002) to conflict by either internalizing negotiations, or when lacking the requisite capacity to resolve disagreements, turn to government authorities.

7.2.2 Agriculture co-management arrangement

The existence of multiple and diverse institutions and actors is a characteristic of the agricultural co-management arrangement around Lake Jipe. This is what other scholars (e.g. Blair, 2001; Benjamin, 2004) call 'institutional pluralism' wherein multiple state, NGO and community actors interact and collaborate in decision making processes and the state is more or less a broker or a referee to ensure that rules of the game are observed by multiple actors in their interactions. Actors and institutions within agricultural co-management arrangement have developed novel ways of collaborating and interacting in addressing the management and sustainable use of the diverse and dynamic natural resources (cf. Armitage, 2008). A key feature of natural resource use in agriculture around Lake Jipe is a geographical upstream and downstream divide, creating a disjuncture between the source and site of resource degradation. To overcome this divide the institutions and actors around the Lake have developed the capacity to fulfil multiple roles within diverse spatial and temporal socio-ecological interactions of resources and resource users, by locating themselves across a diverse range of agricultural practices, resource flows and property regimes.

Three types of institutions – namely co-management institutions, informal community institutions and formal governmental institutions – collaborate and interact in the management of natural resources for agricultural production. The agricultural co-management institutions that have emerged around Lake Jipe (e.g. land councils) are hybrids of formal and customary rules aimed to mediate resource use conflicts when formal and informal rules independently fail to address the conflicts adequately. Similar to what Naguran (2002) and Carter (2008) have found elsewhere, the informal institutions covering agricultural resources around Jipe have evolved from often implicit informal community organizations and practices which entail mechanisms to handle user conflicts on competition for scarce resources. The co-management institutions governing resources around Lake Jipe have made these informal practices more explicit or legible to formal institutions, such as formal laws, rules, regulations, and principles formulated by governmental authorities and used to govern the use of specific natural resources (e.g. land, water, forests).

Apart from governmental and community actors, the agricultural co-management arrangements around Lake Jipe has been facilitated and supported by MIFIPRO, an NGO (actor) who, not restricted by political and geographic boundaries, has been able to play various roles across multiple spatial and temporal scales. The role of MIFIPRO at Lake Jipe supports the wider understanding of the catalytic role of NGOs in the development of co-management arrangements, including capacity building, creating new partnerships and advocating the rights of marginalised groups (cf. McConney *et al.*, 2004; Thompson *et al.*, 1999; Umar & Kankiya, 2004). MIFIPRO in particular has brought much needed flexibility to the co-management of agriculture around Lake Jipe. At one moment the NGO indirectly resolves conflicts between and among resource users separated by different spatial and geographic positions by assisting with improved and more efficient exploitation and management of natural resources. At other, the NGO forges cross-scale linkages between governmental and community actors at different levels through building their capacity in comprehending and enforcing sustainable resource management rules. MIFIPRO therefore both links and empowers community and the governmental actors for sustainable management of land and water resources for agricultural production, and fulfils a crucial role in the establishment and function of cross-scale environmental co-management institutions.

7.2.3 Livestock co-management arrangement

Pastoral land is a competitive resource that is scarce and spatially distributed, but also socially and politically distributed among villages at Lake Jipe social-ecological system. Two groups of users, *Pare* and *Maasai* herders, with different behaviours and cultural backgrounds compete for this resource. The *Pare* regards themselves as pioneers in opening up the land around Lake Jipe for pastoral grazing while regarding the *Maasai* as intruders, with no right to access and use the area. These two groups are governed by diverse and mutually hostile informal institutions. Whereas the *Pare* ethnic group is governed by informal reciprocal norms and traditions, the *Maasai* ethnic group is governed by nomadic norms. However, the temporal and spatial scarcity of pastoral land in the wider region around Jipe has brought these two ethnic groups together in their search for the better-off pasture conditions.

The conflicting claims to pastoral rights around Lake Jipe illustrate similar patterns of conflict to fisheries and agricultural. However, similar to what Kanyongolo (2005) notes with reference to Eastern Africa, the renegotiation of communal access rights to resources can transform the social setting into an arena of conflict between the haves and have-nots. Because of a range of ethnic politics and the formalization of resource boundaries the two pastoral groups around Lake Jipe have increasing come into conflict. The grazing of nomadic *Maasai* herders, unable to formalize their claims over the land in and around *Pare* villages has resulted in tensions and violent clashes between the two ethnic herder groups. What were once dynamic seasonal and spatial claims over land, negotiated through equally dynamic social relations (Cornwall, 2000), have been gradually fossilized thereby excluding nomadic claims.

The emergence of livestock co-management arrangements, combining formal and informal actors and institutions has been instrumental in addressing the conflicts between the two ethnic groups. Interventions by governmental actors using the authoritative power of formal institutions, such as property rights, has contributed to resolving conflicts by enforcing mutual grazing rights of livestock for the two groups within the same spatial extent within designated time periods. Nevertheless, the government indirectly empowers the *Pare,* by asking the *Maasai* to pay a user fee when grazing around the Lake (cf. Sekhar, 2004). The choice of a fee reifies control by the Pare over the Maasai and indicates the continued challenges of seeking what Kideghesho and Mtoni

(2008) refer to social cohesion in resource management, the lack of which dilutes attempts to create durable institutions and, ultimately, *co*-management arrangements. To overcome this challenge of mediating tensions between the two groups of pastoralists, the government has enrolled the assistance of the (informal) elders who use their normative legitimacy, especially within the *Pare* community, to ensure that any conflict does not escalate. At the same time, the elders have become an important institution by facilitating information flows to the district government when conflicts emerge between two herder groups.

7.3 Complexity in analysing co-management arrangements

It has been conventional in co-management studies to analyze social-ecological systems monolithically (e.g. Berkes & Folke, 1998; Carlsson & Berkes, 2003), or to analyze one resource component of these systems (e.g. Nielsen *et al.*, 2004; Thomson & Gray, 2009). But social-ecological systems are often comprised of multiple and diverse but interdependent natural resource management systems. Hence such monolithic analyses overlook the internal complexities and interactions within individual natural resources management systems, and inter- and cross-scale interactions and complexities within one social-ecological system. These internal complexities and interactions influence individual natural resources management systems and the management of the social-ecological system as a whole. Thus viewing a social-ecological system as monolithic (in whatever way) conceals how one natural resources management institution or system imposes conditions on and/or receives incentives from other natural resources management institutions and systems existing within the same social-ecological system. This section reflects on the intra- and inter-sectoral linkages of multiple co-management arrangements and their relevance to natural resource management in Tanzania. As such it integrates on a higher level the findings of the three – partly monolithic – case studies performed at Lake Jipe socio-ecological system.

7.3.1 Adding complexity I: unpacking scales and actors

Through the analysis of emerging co-management at the Lake Jipe socio-ecological system it has become clear that a full understanding of these decentralized and participatory co-management practices do not readily allow for reductionism. Recognizing this allows us to move beyond an understanding of co-management as the vertical interplay between government and community. The Lake Jipe case study clearly demonstrates that both government and community have to be unpacked in at least two ways to really understand the nature of co-management and its wider potential.

First, following Adger *et al.* (2005), Berkes and Seixas (2004) and Cash *et al.* (2006), the notion of scale has to be further elaborated. Co-management involves multiple scales of actors and institutions and cannot be reduced to an analysis of local natural resource intrusions and the local mechanisms that aim to manage those intrusions. Governmental actors and institutions of multiple scales were involved in the analysis, from local village levels, via district and national levels to cross border interactions and actors. This was also the case to a lesser extent with community actors and institutions, where non-resident fisherman and migrating herders use local natural resources.

Secondly and partly related to the former issue of multiple scales, the categories of government and community need to be unpacked. Multiple actors perform formal and informal or hybrid co-management functions. It is unrealistic to lump these actors into just two categories: community and government. As many studies have emphasized (e.g. Armitage *et al.,* 2008; Berkes, 2002), within the community actors are not one but differentiated based on diverse cultural backgrounds, spatial origins (residents and non-residents), and different geographical positions (upstream and downstream resource users), different interests (herders versus agricultural farmers) etc. Similarly, governmental actors are differentiated based on their sectoral specialization into actors that enforce compliance with fisheries, land, and water regulations. And, of course, across multiple scales governmental actors differ in their abilities to govern, their interests and their involvement in institutions. Lumping together these actors as one community and one governmental actor conceals (and therefore ignores) not only internal horizontal politics and intra-scale and cross-sectoral interactions, but also vertical cross scale interactions and politics existing among and between multiple formal/governmental actors and multiple informal/community actors.

This unpacking has direct consequences for the analysis of natural resource co-management. It contributes to new insights, and a new conceptualization, emphasizing the co-existence of multiple institutions, within which different actors can be involved in multiple natural resource management institutions of different kinds: formal, informal and co-management (Figure 7.1). This thesis therefore opens up new conceptual space by distinguishing institutions and arrangements in co-management.

The existence of multiple, diverse and interacting actors and institutions imply the complex characteristic of the Jipe social-ecological system. This complexity manifests itself as the dynamic heterogeneous participation of actors and institutions in multiple and diverse roles of harmonizing resource use in a system with multiple rights and entitlements, and in mediating relational interactions among the respective actors. Whereas congruency of ideas, values and expectations emerge in some cases, in other cases tensions emerge and therefore institutions and actors adopt mediating roles. Both direct and indirect mechanisms are used in mediating these conflicts. Whereas direct mechanisms entail negotiation, deliberation, sanctioning and application of rules that enforce entitlements and rights of various users, indirect mechanisms include empowerment of actors through improvement of their livelihood practices in keeping with efficient and sustainable natural resources management. The hope of doing so is to improve the availability and accessibility of environmental resources and services among competing resource users.

However, some complex co-management arrangements of Jipe social-ecological system have shortcomings. As earlier revealed, sometime formal actors use their authoritative resources illicitly to play contrary to the rules of game to which they should be held accountable and which they should comply with. This implies that recognition as formal actor does not guarantee abidance to formal rules of the game, and other incentive mechanisms are then necessary to ensure that formal actors abide to formal rules of the game. The present co-management arrangements, nonetheless, lack such formal incentive mechanisms for these state actors. Unless such mechanisms are in place, there is a danger of continuing acting against the rules by state actors, while they pretend to abide to the rules.

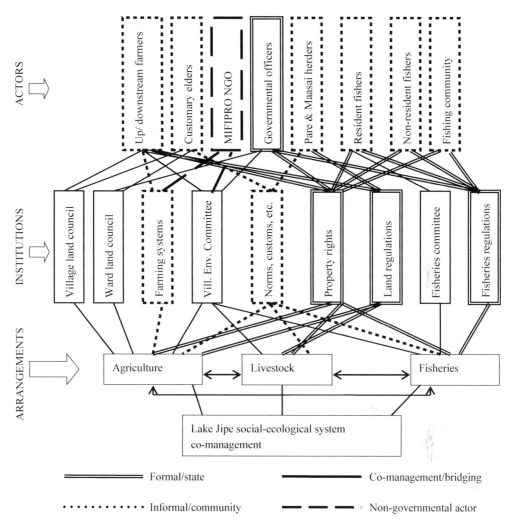

Figure 7.1. Overlapping and layering of institutions and actors in the three co-management arrangements at Lake Jipe social-ecological system.

7.3.2 Adding complexity II: cross-sectoral analyses

Partly in line with the monolithic view of social-ecological systems, this thesis has disaggregated the Lake Jipe social-ecological system into three different natural resources management systems, referred to as co-management arrangements, i.e. fisheries, agriculture and livestock co-management arrangements. In each of these arrangements we have analyzed the interplay of governmental and community actors and of formal and informal institutions (Figure 7.1). Like much of the literature on social-ecological systems (Adger *et al.*, 2005; Berkes, 2003; Berkes & Seixas, 2004; Carlsson & Berkes, 2003; Kendrick, 2003), we see that the three co-management arrangements are different.

However, the interrelations between these arrangements occur at a sectoral level, thereby allowing us to better analyse the complexity inherent in social-ecological systems.

Throughout the analysis multiple reasons for relating these three co-management arrangements have emerged. For one, there are various natural resource interdependencies in Lake Jipe social-ecological system. The practices of actors in one arrangement at one spatial locality (e.g. improper farming practices on the upstream area) and temporal space for example, have implications and/ or can undermine practices and sustainability of another arrangement in another spatial locality at the same or different temporal space. For example, siltation of the Lake negatively influences (future) fisheries productivity. And livestock and agricultural farmers compete for similar natural resources at similar spatial and temporal settings. Complexities are further added in terms of the presence of water as a mobile natural resource flowing from markedly different (upstream and downstream) ecosystems, with multiple actors at multiple locations depending on this flowing natural resource. This does not only have ecological impacts, but also involves socio-institutional linkages at different locations within the catchment of Lake Jipe. It illustrates that disturbances imposed in one spatial position(s) may result in disharmony, disturbance, disruption and/or destabilisation of an entire co-management arrangement or even a socio-ecological system (beyond that sectoral arrangement). These natural resource interdependencies force us to analyse not just internal interactions, linkages and conflicts within single natural resources management systems at one place, but also to compare and contrast inter- and cross-sectoral complexities and interactions at multiple locations among the institutions and actors in comprising these three natural resources management systems at Lake Jipe social-ecological system.

Secondly, the actors and institutions involved in sectoral co-management arrangements cross borders. Lake Jipe social-ecological system comprises multiple arrangements, institutions and actors. Some community, governmental and resource use actors are limited to one co-management arrangement, but several others are not. Similarly, some of the (governmental or community based) institutions in the three co-management arrangements are confined to specific sectoral boundaries, whereas other institutions operate simultaneously at intra- and cross-sectoral levels, beyond the sectoral boundaries. These trans-sectoral boundary institutions govern similar functions within the three distinguished co-management arrangements, and bridge and mediate actors at the interface of these different co-management arrangements. For example, informal institutions (e.g. customary norms) govern the mediation of conflicts among farmers or among livestock keepers, but also conflicts between farmers and livestock keepers when interests of the first resource user group are compromised by the practices and interests of the other resource user group. Hence, institutions – whether formal, informal or co-management – are often not restricted to one sectoral co-management arrangement.

By making the social and ecological interrelations around Lake Jipe legible neither the hierarchy nor the effectiveness of the institutions in governing the practices of actors in the management process have been determined. Rather, I have identified formal and informal institutions that interact and collaborate in the management of natural resources for the three co-management arrangements in Jipe social-ecological system. In doing so this thesis highlights the 'layering' of co-management (Figure 7.1). Using this geographical metaphor, it is possible to structure the interactions between actors, institutions and arrangements in natural resources co-management in a socio-ecological system. However, as already noted, these layers are not hierarchical but instead networked, composed of actors, institutions and arrangements that created relational inter-dependencies. This layering concept adds new understanding on the existing theory on co-management of social-ecological system, which as noted above has tended to become too monolithic (in various ways).

Therefore, this thesis adds to the existing theory with the argument that governmental-community linkages are not simplistic (or monolithic) but are a complex layering of multiple governmental-community entities and institutions in multiple co-management arrangements. This challenges and adds to the current concept of the existence of one co-management arrangement in one social-ecological environment (Pinkerton, 1994; Pomeroy, 1995; Pomeroy *et al.*, 1995; Pinto da Silva, 2004; Singleton, 2000). I propose to label this perspective (as it is far too early to call it a theory) the complex layering of co-management.

7.3.3 Relevance for natural resources management in Tanzania and East Africa

This thesis has unveiled the shift from classic to populist natural resources management regimes in Tanzania. The centralized natural resources management regimes have come to the point that they cannot address the challenges of mitigating the degradation of natural resources due, in part, to the government's incapacity in terms of human and financial resources to monitor and control unsustainable natural resources management regimes and practices. Recognising the potential for intra and inter sectoral interrelations and dependencies, co-management offers a new opportunity for governmental and community institutions and actors to collaborate in the management of natural resources in complex social-ecological systems such as Lake Jipe. However, similar to the findings of Pomeroy *et al.* (2001) and others, a key reason for the ongoing success of such these –co-management arrangements may also be the space left for independent community and governmental institutions. Hence, for co-management in Lake Jipe and Tanzania to continue, both formal legal systems with their authoritative enforcement and informal community institutions such as elders need to co-exist, both contributing to natural resource management in a wider set of co-management arrangements.

Successful co-management arrangements also require the formal integration of the community actors and institutions in the management of natural resources in Tanzania. Such integration is not a new phenomenon. As has been noted in chapter 3, during the colonial times the colonial government integrated chiefdom-based institutions in the management of natural resources in Tanzania (then Tanganyika). During that time, the local chiefs were given authority to enforce compliance with governmental rules over the sustainable use and management of natural resources. The integration of the local people in the management of natural resources nonetheless was undermined and disrupted during the independence and post-independence era during which the Tanzanian government, like other governments in East Africa, terminated the participation of the local institutions. This thesis has demonstrated that the reintegration of local institutions in the management of natural resources at Lake Jipe by the Tanzanian government is evident and shows promise for the sustainable natural resource management.

Despite the recognition given to vertically integrating formal and informal natural resources management institutions in Tanzania less attention has been given to understanding sectoral cross-linkages. In other words, management (and research) has been largely confined to the intra-sectoral relationships, ignoring inter- and cross-sectoral interactions of natural resources management systems. As the case of Lake Jipe clearly illustrates, natural resources management systems in Tanzania are comprised of complex physical and institutional inter-relationships and inter-dependences. The results from Lake Jipe therefore indicate the restoration or revival of local institutions in the management of natural resources, as well as demonstrating that multiple natural resources in one social-ecological environment are interdependent and actors and institutions governing them cross sectoral natural resource boundaries.

7.4 Recommendations

In this section the thesis question 'what possibility is there for the government and the community to sustainably co-manage Lake Jipe wetland in Tanzania' is answered. As there exist co-management arrangements comprised of governmental and local community actors, within multiple formal and informal institutions, the basic conditions for strengthening co-management of Lake Jipe wetland are present.

For sustainable co-management of the Lake Jipe social-ecological system, the following challenges or weaknesses need to be resolved. First, there are deficient capacity building mechanisms. Presently, although some empowerment initiatives are given to the local stakeholders, such programmes are limited to a few local actors, while the majority is not served. For example, although fishers are generally accused of engaging unsustainable fishing tools in fishing, the government has given training on sustainable fishing practices to some fishers who are members in fisheries committees. But those outside the committees and especially non-resident fisherman have been deprived of such trainings. By the same token, although the government promotes formation of co-management institutions for sustainable environmental management, it was through the Non-Governmental Organisation MIFIPRO that these institutions were relatively empowered because of among others the alleged financial inadequacy of the government. Hence there is a need for governmental authorities to increase their contribution in preparing communities and natural resource users to become involved in co-management.

Second, although some cross-sectoral informal institutions are linked to different forms of co-management, there is a limited support to ensure that resource user-groups are motivated to comply with these informal institutions. While for some cases there has been formal governmental support to enforce the compliance with informal institutional mechanisms, for other cases such mechanisms are missing. The compliance with significant informal institutions could be enhanced through the enforcement wherein the government would, for example, raise the sanctions for non-compliance when violators are referred to the government. This would encourage and enhance mediation at the community level and would imply less involvement of the government as a mediator/facilitator.

Third, some alternative ways of securing socio-economic interests are imperative for sustainable natural resources management at Lake Jipe socio-ecological system. Presently there is a high tension between livelihoods and environmental interests. This pressure exacerbates with increasing uncertainties of meeting socio-economic needs. Unless new alternative ways of meeting such needs are integrated and/or the existing ones are improved there will be a limited possibility for maintaining sustainable co-management of Lake Jipe social-ecological system.

Finally, based on the wider and diverse nature of social-ecological systems socio-economically, culturally and ecologically in Tanzanian and East African context, further analysis on institutional and actors interplay entailing the community and governmental levels or even including other levels beyond the government and the community is recommended. One potential area for further analysis can be on the effectiveness of the various informal, formal and co-management institutions of the studied multiple co-management arrangements for sustainable management of social-ecological systems, in order to provide concrete conclusions and recommendations on the co-management of complex social-ecological systems in Tanzania and East Africa. One can expect that a number of the findings of this study on Lake Jipe will have relevance for other wetlands in Tanzania (and other East African countries) and even for other socio-ecological systems within the country, further research needs to prove that. The road to better understanding co-management is still being sought; but the conceptual contributions made in this thesis provide a step in the right direction.

References

Adger, W.N. (1999). Social Vulnerability to Climate Change and Extremes in Coastal Vietnam. *World Development* 27(2): 249-269.

Adger, W.N., Brown, C., & Tompkins, E.L. (2005). The Political Economy of Cross-Scale Networks in Resource Co-Management. *Ecology and Society* 10(2): 9. Retrieved from http://www.ecologyandsociety.org/vol10/iss2/art9/.

Adger, W.N., Benjamisen, T., Brown, A.K., & Svarstad, H. (2001). Advancing Political Ecology of Community in Natural Resources Conservation. *World Development* 27: 629-49.

Adjewodah, P., & Beier, P. (2004). Working with traditional authorities to conserve nature in West Africa. *African Conservation Telegraph* 1(2). Retrieved from http://www2.for.nau.edu/research/pb1/Service/Adjewodah-Beier_Traditional_Authority_Conservation.pdf.

Agnew, J. (1997). The dramaturgy of horizons: geographical scale in the Reconstruction of Italy by the new Italian political parties, 1992-95. *Political Geography* 16(2): 99-121.

Agrawal, A. (2001). Small Is Beautiful, but Is Larger Better? Forest-Management institutions in the Kumaon Himalaya, India. In: Clark, C., Margaret, A. & Ostrom, E. (Eds.), *People and Forests, Communities, Institutions, and Governance*. The MIT Press. Cambridge, Massachusetts, London, England. Pp 27-55.

Agarwal, B. (2001). Participatory exclusions, community forestry, and gender: an analysis for South Asia and a conceptual framework. *World Development* 29, 1623-1648.

Ahmed, M., Capistrano, A.D., & Hossain, M. (1997). Experience of partnership models for the co-management of Bangladesh fisheries. *Fisheries Management and Ecology* 4: 233-248.

Armitage, D. (2005). Adaptive Capacity and Community-Based Natural Resource Management. *Environmental Management* 35(6): 703-715.

Armitage, D., Marschke, M., and Plummer, R. (2008). Adaptive co-management and the paradox of learning. *Global Environmental Change* 18: 86-98. Retrieved 4th June 2009, from www.sciencedirect.com.

Ascher, W. (1995). *Communities and Sustainable Forestry in Developing Countries*. San Francisco: ISC Press. 40 pp.

Assens, J., & Jensen, K. (Eds.) (2003). *Profile of labour market and trade unions in Tanzania*. LO/FTF council. Danish Trade Union council for international development cooperation. Retrieved from http://www.ulandssekretariatet.dk/files/oplysning/Tanzania%20report.pdf.

Baker, J., Wallewik, H., Obama, J., & Sola, N. (2002). *The Local Government Reform Process in Tanzania: Towards a greater interdependency between local government and civil society at the local level?* Research and Development report no. 6/2002, 67 pp. Agderforskning. Kristiansand. Norway.

Baland, JM., & Platteau, JP. (1996). *Halting degradation of natural resources: Is there a role for rural communities?* FAO of the United Nations and Oxford University Press, Oxford, UK. 423pp.

Baldus, R., Benson, K., & Siege, L. (2003). Seeking conservation partnerships in the Selous Game Reserve, Tanzania Conservation Partnerships in Africa. *Parks* 13(1): 50-80.

Bangs, E., & Shivik, J. (2001). Managing wolf conflict with livestock in the Northwestern United States. *Carnivore Damage Prevention News* No. 3: 2-5.

Barbier, E.B. (1997). The economic determinants of land degradation in developing countries. *Philosophical Transactions Royal Society* London B 352: 891-899.

Barrett, C.B., Brandon, K., Gibson, C. & Gjertsen, H. (2001). Conserving tropical biodiversity amid weak institutions. *BioScience* 51: 497-502.

Barrow, E., & Mlenge, W. (2003). Trees as key to pastoralist risk management in semi-arid landscapes in Shinyanga, Tanzania and Turkana, Kenya. The International Conference on Rural Livelihoods, Forests and Biodiversity 19-23 May 2003, Bonn, Germany. Retrieved from http://www.cifor.cgiar.org/publications/corporate/cd-roms/bonn-proc/pdfs/papers/T3_FINAL_Barrow.pdf.

Belsky J. (1999). Misrepresenting communities: the politics of community-based rural ecotourism in Gales Point Manatee, Belize. *Rural Sociology* 64: 641-667.

Benjamin, C.E. (2004). *Livelihoods and institutional development in the Malien Sahel: A political economy of decentralised natural resource management*. PhD thesis, the University of Michigan, 399 pp. Retrieved from www.williams.edu/CES/ces/people/cbenjamin/dissertation.pdf.

Berkes, F. (2000). *Cross-Scale Institutional Linkages: Perspectives from the Bottom Up*. NAS/NRC Panel on Institutions for Managing the Commons. IASCP 2000 Conference, Indiana University, June 2000. Retrieved from http://dlcvm.dlib.indiana.edu/archive/00000217/00/berkesf041200.pdf.

Berkes, F. (2003). *Can Cross-Scale Linkages Increase the Resilience of Social-Ecological Systems?* Draft paper. RCSD International Conference, Politics of the Commons, Chiang Mai, July 2003. Retrieved from http://dlcvm.dlib.indiana.edu/archive/00001097/00/Fikret_Berkes.pdf.

Berkes, F., and Seixas, C.S. (2004). Building Resilience in Lagoon Social–Ecological Systems: A Local-level Perspective. *Ecosystems* 8: 967-974.

Berkes, F. (2002). Cross-Scale Institutional Linkages: Perspective from the Bottom Up, In: Ostrom, E., Dietz, T., Dolšak, N., Stern, PC. Stonich, S. & Weber, E U. (Eds.), *The Drama of the Commons*. Washington, DC: National Academy Press, pp. 293-321.

Berkes, F & Folke, C. (Eds.) (1998). *Linking social and ecological systems. Management practices and social mechanisms for building resilience*. Cambridge University Press. 459 pp.

Biot, Y., Blaikie, P.M., Jackson, C., & Palmer, J.R. (1995). *Rethinking research on land degradation in developing countries*. The International Bank for Reconstruction and Development. The World Bank Discussion Papers 289. 139pp.

Blaikie, P. (1996). *New Knowledge and Rural Development: a review of views and practicalities*. A paper for the 28th International Geographical Congress Hague, 5th -10th August.

Blaikie, P., Brown, K., Stocking, M., Tang, L. Dixon, P., & Sillitoe, P. (1997). Knowledge in Action: Local Knowledge as a Development Resource and Barriers to its Incorporation in Natural Resource Research and Development. *Agricultural Systems* 55(2): 217-237.

Blair, H. (2001). Institutional pluralism in public administration and politics: Applications in Bolivia and beyond. *Public Administration and Development* 21: 119-129.

Bootsma, H.A., & Hecky, R.E. (1993). Conservation of the African Great Lakes. A limnological perspective. *Conservation Biology* 7, 644-656.

Borrini-Feyerabend, G., Farvar, M.T., Nguinguiri, J.C. & Ndangang, V. (2000). *Co-management of Natural Resources: Organizing Negotiation and Learning by Doing*. Heidelberg, Ge: Kasparek Verlag. Retrieved from http://learningforsustainability.net/pubs/cmnr/cmnr.html.

Bromley, D. (1992). The commons, property, and common-property regimes. In D. Bromley (ed.), *Making the commons work: Theory, practice, and policy*, pp 3-15. San Francisco, CA: Institute for Contemporary Studies Press.

Brown, D. (2003). Is the best the enemy of the good? Livelihoods perspectives on bushmeat harvesting and trade – some issues and challenges. The International Conference on Rural Livelihoods, Forests and Biodiversity 19-23 May 2003, Bonn, Germany. Retrieved from http://www.cifor.cgiar.org/publications/corporate/cd-roms/bonn-proc/pdfs/papers/T2_FINAL_Brown.pdf.

Brown, K. (2003). Three Challenges for a Real People-Centred Conservation. *Global Ecology and Biogeography* (12)2: 89-96.

Buttel, F.H. (1998). Some observations on states, world orders, and the politics of sustainability. *Organization and Environment* 11, 261-286.

Campbell, D.J. (1981). Land use competition at the margins of the rangelands: an issue in development strategies for semi-arid areas. *African Journal of Ecology* 47(1): 55-61.

Campbell, D.J., Misana, S.B., & Olson, J.M. (2004). *Comparing the Kenyan and Tanzanian slopes of Mt. Kilimanjaro: Why are the neighbouring land uses so different? The Land Use Change, Impacts and Dynamics Project.* Working Paper No. 44. Retrieved from http://www.lucideastafrica.org/publications/Campbell_LUCID_WP_44.pdf.

Carlsson, L., & Berkes, F. (2003). *Co-management across levels of organization: concepts and methodological implications.* Lead paper prepared for the Resilience panel at the Regional Workshop of The International Association for the Study of Common Property (IASCP), Politics of the Commons: Articulating Development and Strengthening Local Practices, Chiang Mai, Thailand, July 11-14, 2003. Retrieved from http://dlcvm.dlib.indiana.edu/archive/00001133/00/Lars_Carlsson.pdf.

Carter, S. (2008). Research and Policy Implementation for More Equitable and Sustainable Use of Common-Pool Resources. *Society & Natural Resources* 21(2):152-159.

Cash, D.W., Adger, W., Berkes, F., Garden, P., Lebel, L., Olsson, P. et al. (2006). Scale and cross-scale dynamics: governance and information in a multilevel world. *Ecology and Society* 11(2): 8. Retrieved from http://www.ecologyandsociety.org/vol11/iss2/art8/.

Castello, L. (2001). *Stock Assessment and Management of the Arapaima in the North Rupununi, Guyana.* Instituto de Desenvolvimento Sustentável Mamirauá. 29 pp. Retrieved from http://www.iwokrama.org/library/pdfdownload/Arapaima%20Stock%20Assessment%20and%20Management.pdf.

Castro, A.P., & Nielsen, E. (2001). Indigenous people and co-management: implications for conflict management. *Environmental Science and Policy* 4: 229-239.

Cheong, SM. (2004). Korean fishing communities in transition: limitations of community-based resource management. *Environment and Planning* 37: 1277-1290.

Christiansson, C. (1988). Degradation and Rehabilitation of Agro-pastoral Land: Perspectives on Environmental Change in Semiarid Tanzania. *Ambio*, Vol. 17, No. 2, Ecosystem Redevelopment, pp. 144-152.

Christie, P., Buhat, D., Garces,LR. & White, AT. (2003a). The challenges and rewards of community- based coastal resources management. In: Brechin, S.R., Wilshusen, P.R., Fortwangler, C.L. and West, P.C. (Eds.), *Contested nature, promoting international biodiversity with social justice in the twenty-first century.* State University of New York Press, Albany. pp. 231-249.

Cornwall, A. (2000). *Making a difference? Gender and participatory Development.* IDS discussion paper 378. Sussex: Institute of Development Studies, 39pp, Retrieved from http://www.unidaddeigualdad.es/documentos_contenidos/1202_59101840_MakingADifference.pdf.

Dadzie, S. & Haller, R.D. (1988). A note on the fishes of Lake Jipe and Lake Chale on the Kenya-Tanzanian Border. *Journal of East Africa Natural History Society and National Museums of Kenya* 192 46-52.

Douglas, M. (1992). *Risk and Blame: Essays in Cultural Theory.* London/New York: Routledge. 324 pp.

Downie, D. & Fenge, T. (Eds). (2003). *Northern Lights Against POPs: Combating Toxic Threats in the Arctic.* McGill-Queen's University Press.

Ehrlich, P., & Ehrlich, A. (1991). *The Population Explosion.* New York: Touchstone, Simon and Schuster Inc.

Elmqvist, T. (2000). *Indigenous Institutions, Resilience and Failure of Co-Management of Rain Forest Preserves in Samoa*. Presented at Constituting the Commons: Crafting Sustainable Commons in the New Millennium, the 8th Conference of the International Association for the Study of Common Property, Bloomington, Indiana, USA, May 31-June 4. Retrieved from http://74.125.155.132/scholar?q=cache:WzEEAo84KaAJ:scholar.google. com/+Indigenous+Institutions,+Resilience+and+Failure+of+Co-Management+of+Rain+Forest+Preserves+ in+Samoa.&hl=nl.

Environmentalists Sans Frontieres (ESF) (2005). *Baseline survey report for lake Jipe. Emergency community-based recharge interventions for the restoration of lake Jipe*. 125 pp. Retrieved from www.esfconsultants.org.

Fajber, E. (2005). Participatory research and development in natural resource management: Towards social and gender equity. In J. Gonsalves, T. Becker, A. Braun, D. Campilan, H. de Chavez, E. Fajber, et al. (Eds.), *Participatory research and development for sustainable agricultural and natural resources management*. A source Book. International Development Research Centre, South Asia Regional Office. Retrieved from http://www.idrc.ca/ en/ev-85048-201-1-DO_TOPIC

Feeny, D., Berkes, F., McCay, B.J., & Acheson, J.M. (1990). The tragedy of common: Twenty two years later. *Human Ecology* 1, 1-19.

Feeny, D. (1988). The demand for and supply of institutional arrangements. In Ostrom, V., Feenly, D. & Picht, H. (Eds), *Rethinking Institutional Analysis and Development*, San Francisco: International Centre for Economic Growth, Chapter 6, pp.159-209.

Figueroa, M. (2002). Co-management and valuation of Caribbean coral reefs: A Jamaican NGO perspective. World Fish Centre. 182-191. Retrieved from http://cermes.cavehill.uwi.edu/mpa/NMP_evaluation_report.pdf.

Food and Agriculture Organisation (FAO) (1998) Wetland Characterization and classification for sustainable agricultural development. Retrieved from http://www.fao.org/docrep/003/x6611e/x6611e03.htm#P9_84

Fresco, L.O.; Hoogeveen, H.; Burke, J.; Graaf, J.; Halsema, G.E.; Hermans, L.M.; et al. (2005). Water for Food and Ecosystems: the road from resource dispute to reconciliation. *Sustainable Development International* 13 (08/1): 86-89.

Fritz, H., & Duncan, P. (1994). Carrying Capacity for Large Ungulates of African Savanna Ecosystems. *Proceedings of Biological Sciences*, 256 (1345): 77-82, April 22, 1994, The Royal Society Stable. Retrived from http://www. jstor.org/stable/49597.

Gaudet, J.J. (1975). Report on the Aquatic Weed Problems at Lake Jipe (Tanzania), USAID (REDSO) Nairobi.

Geheb, K. & Crean, K. (2000). The Co-management Survey Report: an executive summary. In K. Geheb, & K. Crean (Eds.), *The Co-management Survey. Co-managerial Perspectives for Lake Victoria Fisheries*. Lake Victoria Fisheries Research Project Technical Document No 11 LVFRP/TECH/00/11. The Social Economic Data Working Group of Lake Victoria Fisheries Research Project, Jinja, pp. 1-10

German, L., Shenkut, A., & Zenebe, A. (2008). Managing Linkages Between Communal Rangelands and Private Cropland in the Highlands of Eastern Africa: Contributions to Participatory Integrated Watershed Management. *Society and Natural Resources* 21(2):134-151.

Gibbons, M. (1999). Science's new social contract with society. *Nature* 402(81): 11-17.

Giger, M. (1999). *Avoiding the shortcut: Moving beyond the use of direct incentives. A review of the experience with the use of incentives in projects for sustainable soil management*. Development and Environment Reports No 17. Retrieved from http://www.cde.unibe.ch/Themes/pdf/incentives_e.pdf.

Girot, P.O., Weitzner, V. & Borrás, M.F. (1998). From conflict to collaboration: The case of Cahuita National Park, Limón, Costa Rica. Retrieved from http://www.iascp.org/Drafts/fonseca.pdf.

Glaser, M., & da Silva, O.R. (2004). Prospects for the co-management of mangrove ecosystems on the North Brazilian coast: Whose rights, whose duties and whose priorities? *Natural Resources Forum* 28: 224-233.

Grinnel, R.M. (2001). Social work research and evaluation: Quantitative and qualitative Approaches (6th ed). Itasca, IL. Peacock. In A.U. Yvonne, A.P. Gabor, & R.M. Grinnell (Eds.), *Evaluation in social work: the art and science of practice.* pp 367-381.

Guha, R. (1997). The authoritarian biologist and the arrogance of anti-humanism: wildlife conservation in the Third World, *The Ecologist,* 27(1), 14-20.

Gwebu, T. (2001). Localized Wood Resource Depletion in Botswana: Towards a Demographic, Institutional and Cosmovisional Explanation. *Ethics, Place and Environment* 5(2):144-152.

Hachongela, P., Jackson, J., Malasha, I., & Sen, S. (1998). Analysis of emerging co-management arrangements: Zambian inshore fisheries of Lake Kariba. In Fisheries co-management in Africa. In: Normann, A., Nielsen, J. and Sverdrup-Jensen, S. (Eds.), *Proceedings from a regional workshop on fisheries co-management research,* pp. 79-98. Hirtshals, Denmark: IFM.

Hara, M., Donda, S. & Njaya, F.J. (2002). Lessons from Malawi's experience with fisheries co-management initiatives. In: Geheb, K. and Sarch, M. (Eds.), *Africa's inland fisheries: The management challenge, pp.* 31-48. Kampala: Fountain Publishers.

Hardin, G. (1968). The tragedy of the commons. *Science* 162(3859): 1243-1248.

Harrill, J.C. (2002). Marine wetland interactions and policy in Tanzania. Retrieved from http://iodeweb1.vliz.be/ odin/bitstream/1834/533/1/Wetlands125137.pdf.

Hellin, J. & Schrader, K. (2003). The case against direct incentives and the search for alternative approaches to better land management in Central America. *Agriculture, ecosystems & Environment,* 99: 61-81.

Hodgson, D.L. (2000). Taking stock: state control, ethnic identify an pastoralist development in Tanganyika 1948 – 1958. *Journal of African History* 41: 55-78.

Howitt, R. (1998). Scale as relation: musical metaphors of geographical scale. *Area* 30:49-58.

International Centre for Living Aquatic Resources Management (ICLARM) (2001). *Sustainable management of coastal fish stock in Asia.* Final Report of Project ADB RETA 5766 to the Asian Development Bank ICLARM, Penang, Malaysia. 34p.

International Fund for Agricultural Development (IFAD) (1996). *Combating desertification. The role of incentives.* A desk review of literature and practical experiences with incentives in natural resources management in sub-Saharan Africa; with emphasis on gender, institutional and policy dimensions. IFAD and centre for development cooperation services (CDCS) Vrije University Amsterdam.

Isaac, V.J., Ruffino, M.L., & McGrath, D. (1998). The experience of community based management of middle Amazonian Fisheries. 15 pp. Retrieved from http://www.indiana.edu/~iascp/Final/isaac.pdf.

IUCN (2000). *Proceedings of Lake Jipe Cross-Border Planning Workshop,* Nairobi. Eastern Africa Regional Office. Retrieved on 20 July 2007 from http://www.worldwildlife.org/bsp/publications/africa/121/121/refs.htm.

Jackson R., & Wangchuk, R. (2001). Linking Snow Leopard Conservation and People-Wildlife Conflict Resolution: Grassroots Measures to Protect the Endangered Snow. *Endangered Species UPDATE.* 18(4): 138-141.

Jackson, J., Muriritirwa, W., Nyakahadzoi, K., & Sen, S. (1998). Analysis of emerging co-management arrangements in Zimbabwean inshore fisheries of lake Kariba. In A.K. Norman, J.R. Nielsen & J.S. Sverdrup (Eds.). Fisheries Co-Management in Africa. Proceedings of a regional workshop on fisheries co-management research, Mangoshi, Malawi 18–23 March 1997. Fisheries Co-Management Research Project, Research Report No 12 (99-124).

Jentoft, S. (2005). *Property Rights and Co-management in Artisanal Fisheries*. ICSF-CeDePesca seminar on Recognition of Property Rights and Access to Fisheries Resources: Conditions for Sustainable Fisheries in Latin America. Santa Clara del Mar, Argentina, March 1-4, 2005. Retrieved from http://draugen.nfh.uit.no/~arnee/MaReMa/argentina_samudra_artikkel.pdf.

Jentoft, S. (2004). Fisheries co-management as empowerment. *Marine policy* 29 (2005) 1-7. Retrieved from www.elsevier.com/locate/marpol.

Jentoft, S. (1999). *Commons in a Cold Climate. Coastal Fisheries and Reindeer Pastoralism in Northern Norway: the Co-Management Approach*. Parthenon publishers, Paris, pp 17-39.

Jentoft, S., McCay, BJ., & Wilson, D.C. (1998). Social theory and fisheries co-management. *Marine policy* 22(4-5): 423-436.

Jodha, N.S. (1996). Property Rights and Development. In: Hanna, S.S., Folke, C. and Maler, K.G. (Eds.), *Rights to Nature: Ecological, Economic, Cultural, and Political Principles of Institutions for the Environment*, Covelo, CA: Island Press. pp 205-222.

Kajembe, G.C., Nduwamungu, J. & Luoga, E.J. (2006). The impact of community based forest management and joint forest management on the forest resource base and local peoples' livelihoods. Case studies from Tanzania. *Commons Southern Africa Occasional Series*, No. 8. Centre for Applied Social Sciences, University of Harare, Zimbabwe. Retrieved from http://www.sarpn.org.za/documents/d0001454/P1800-Kajembe-et-al_PLAAS_CASS.pdf.

Kajembe, G.C., Monela, G.C., & Mvena, Z.S.K. (2003). Making community management work: A case study of Duru-Haitemba village forest reserve, Babati, Tanzania. Second International Workshop on Participatory Forestry in Africa. Defining the Way Forward: Sustainable Livelihoods and Sustainable Forest Management Through Participatory Forestry. Pp 169-172.

Kanyongolo, F.E. (2005). Land occupations in Malawi: challenging the neoliberal legal order. In Moyo, S. and Yeros, P. (Eds.), *Reclaiming the land: The resurgence of rural movements in Africa, Asia and Latin America*. Pp 118-141.

Kassenga, G.R. (1997). A descriptive assessment of the wetlands of the Lake Victoria basin in Tanzania. *Resources Conservation and Recycling*, 20(2), 127-141.

Katon, B., Pomeroy, R, Salamanca, A. (1997). *The marine conservation project for San Salvador: a case study of fisheries co-management in the Philippines*. Fisheries Co-management Research Project Working Paper no. 23. Manila: ICLARM, 1997. pp. 1-95.

Kauzeni, A.S., Shechambo FC., & Juma, I. (1998). Private and communal property ownership regimes in Tanzania. 11pp. Retrieved from ftp.fao.org/sd/sda/sdaa/LR98_1/art5.pdf.

Kebe, M. (1998). Artisanal fisherfolk's involvement in fisheries rehabilitation in Senegal: Co-management perspectives. In A.K. Norman, J.R. Nielsen, & J.S. Sverdrup (Eds.), *Fisheries Co-Management in Africa*. Proceedings of a regional workshop on fisheries co-management research, Mangoshi, Malawi 18-23 March 1997. Fisheries Co-Management Research Project, Research Report No 12; pp. 291-302.

Kendrick, A. (2003). Caribou Co-management in Northern Canada: Respecting Multiple Ways of Knowing. In F. Berkes, J. Colding, and C. Folke (eds.), *Navigating Social-Ecological Systems: Building Resilience for Complexity and Change* (Cambridge, UK: Cambridge University Press), pp. 241-69.

Kirk, M. (1999). The Context for Livestock and Crop-Livestock Development in Africa. In N. McCarthy, B. Swallow, M. Kirk, & P. Hazell (Eds.), *The Evolving Role of the State in Influencing Property Rights over Grazing Resources in Sub-Saharan Africa*, pp. 23-54.

Kideghesho, J.R. & Mtoni, P.E. (2008). The potentials for co-management approaches in western Serengeti, Tanzania. Mongabay.com Open Access Journal - *Tropical Conservation Science*.1(4):334-358.

Kumar, S. & Kant, S. (2007). Exploded logit modeling of stakeholders' preferences for multiple forest values. *Forest Policy and Economics* 9: 516-526.

Kumar, D. (2005). *Community empowerment for fisheries co-management*. Paper prepared for the APFIC Regional Workshop on Mainstreaming Fisheries Co-management in Asia Pacific – Siem Reap, Cambodia, 9-12 August 2005. 10 pp. Retrieved from ftp://ftp.fao.org/docrep/fao/008/ae940e/ae940e00.pdf.

Leach, M., Mearns, R., & Scoones, I. (1999). Environmental entitlements: Dynamics and institutions in community-based natural resource management. *World Development* 27(2):225-47.

Lebel, L., Garden, P. & Imamura, M. (2005). The politics of scale, position and place in the management of water resources in the Mekong region. *Ecology and Society* 10(2): 1-18. Retrieved from http://www.ecologyandsociety. org/vol10/iss2/art18.

Lélé, S. (2004). Decentralising Governance of Natural Resources in India: A review. Centre for Interdisciplinary Studies in Environment and Development. ISEC, Nagarabhavi, Bangalore 560 072, India. 42pp. Retrieved from http://www.cised.org/wp-content/uploads/dgnr-in-india-final.pdf.

Likens, G.E. (1992). *The ecosystem approach. Its use and abuse*. Oldendorf/Luhe, Germany: Excellence in Ecology, Ecology Institute.

Lim, C.P., Yoshiaki, M. & Yukio, S. (1995). Co-management in Marine Fisheries: The Japanese Experience. *Coastal Management* 23: 195-221.

Lim, L.L. (2006). Facilitating Civil Society-Government partnership in coastal resources management: the case of the Integrated Fisheries and Aquatic Management Council of Northern Lamon Bay, Province of Quezon, Philippines. Retrieved from http://www.sanford.duke.edu/centers/civil/papers/lim.pdf.

Limbu, F.L. & Mashindano, O.J.N. (2002). Agricultural Sector and Poverty in Tanzania. In Mbelle, Mjema & Kilindo (Eds). *The Nyerere Legacy and Economic Policy Making in Tanzania*. Dar es Salaam University Press.

Lingard, M.L., Raharison, N., Rabakonandrianina, E., Rakotoarisoa, J.A. & Elmqvist, T. (2003). The role of local taboos in the conservation and management of species: The Radiated Tortoise in Southern Madagascar. *Conservation and Society* 1(2): 223-246.

Lyatuu, H.A. (1981). A Report on the Survey of Lake Jipe Weed Infestation, Kilimanjaro Region, Tanzania.

Madulu, N.F. (2005). Impacts of Population Pressure and Poverty Alleviation Strategies on Common Property Resource Availability in Rural Tanzania. *African Journal of Environmental Assessment and Management* 10: 26-49. Retrieved July 22, 2009 from http://www.tzonline.org/pdf/impactsofpopulationpressureandpovertyalleviation. pdf.

Maembe, A. (2004). Lake Jipe awareness raising strategy (2005 -2007). Ministry of Natural Resources and Tourism (wildlife division). The United Republic of Tanzania. Retrieved from http://www.ramsar.org/pdf/outreach_ actionplan_tanzania_jipe.pdf.

Maghimbi, S. (1994). Pre-capitalist modes of production in Tanzania: reference to modes of production in the Old Ugweno State of North-eastern Tanzania. *Utafiti* NS 1: 20–35.

Marston, S.A. (2000). The social construction of scale. *Progress in Human Geography* 24: 219-42.

Masija, E.H. (1991). Irrigation of wetland in Tanzania. In Proceedings of a Seminar on Wetlands of Tanzania. Morogoro, Tanzania. pp 73-75.

Masudi, A.S, Mashauri, D.A,& Mayo, A.W. (2001). Constructed wetlands for wastewater treatment in Tanzania. Retrieved from www.scienceinafrica.co.za/2001/september/wsp.htm.

Matthews, G.V.T. (1993). The Ramsar Convention on Wetlands: its history and development. Ramsar Convention Bureau, Gland. 120 pp.

McCay, B., (2002). Emergence of institutions for the commons: contexts, situations, and events. In: Ostrom, E., Dietz, T., Dols´ak, N., Stern, P.C., Stovich, S., Weber, E.U. (Eds.), 2002. *The Drama of the Commons*. National Academy Press, Washington, DC, pp. 361-402.

McConney, P., Pomeroy, R., & Mahon, R. (2004). *Coastal Resources Co-Management in the Caribbean*. Presented at The Commons in an Age of Global Transition: Challenges, Risks and Opportunities, the Tenth Conference of the International Association for the Study of Common Property, Oaxaca, Mexico, August 9-13. Retrieved from http://dlcvm.dlib.indiana.edu/archive/00001439/00/McConney_Coastal_040512_Paper389.pdf.

McNab, D.C. (2004). *Changing livelihood-environment links and opportunities for enhancing livelihoods in the poverty-stricken dry land region of Sekhukhune District in South Africa*, PhD thesis, University of Sheffield, UK.

Meroka, P. (2006). *Common pool resource management and conflict resolution in Rufiji Floodplain,Tanzania*. Unpublished doctoral dissertation. University of Zurich, Department of Social Anthropology.

Meertens, B. (2000). Agricultural performance in Tanzania under structural adjustment programs: Is it really so positive? *Agriculture and Human Values* 17:333-346.

Ministry of Agriculture and Cooperatives (MAC) (1997). *Agricultural and Livestock Policy 1997*. The United Republic of Tanzania.

Ministry of Livestock Development (MLD) (2006). *National Livestock Policy (2006)*. The United Republic of Tanzania. 81 pp.

Ministry of Livestock Development (MLD) (2006). Tamko la Serikali kuhusu kuhamisha mifugo kutoka vyanzo vya maji katika mabonde ya Usangu-Ihefu na Kirombero. Retrieved from www.mifugo.go.tz/documents_storage/ tamko_la_serikali.doc.

Ministry of Natural Resources and Tourism (MNRT) (2003). Sustainable wetland management (2004-2009) in Tanzania. Environmental support programme (ESP). Component document. Ref. No. 104 Tanzanai.1.MFS.23

Ministry of Natural Resources and Tourism (MNRT) (2004). Lake Jipe awareness raising strategy (2005-2007). Ministry of Natural Resources and Tourism, Wildlife Division. Dar es Salaam, Tanzania. 88 pp.

Mitchell, C.R. (1981). *The structure of international conflict*. London: Macmillan. Pp 341.

Mniwasa, E., & Shauri, V. (2001). Review of the Decentralization Process and its Impact on Environmental and Natural Resources Management in Tanzania. Lawyers Environmental Action Team, Dar es Salaam - Tanzania, p.29 (2001) Retrieved from http://www.leat.or.tz/publications/decentralization.

Mtalo, F. (2005). Water resources management issues and conflicts resolution at a catchment level. A case study of Pangani River basin, Tanzania. *FWU, volume 3, topics of integrated watershed management- Proceedings*. Retrieved from http://www.uni-siegen.de/fb10/fwu/ww/publikationen/volume0305/pdf/mtalo.pdf.

Mwarabu, A.K.O. (2007). Taarifa ya uchunguzi wa uhamishaji wa wafugaji toka Mbarali Mbeya kwenda mikoa ya Lindi na Pwani iliyofanywa na timu ya wataalam tarehe 21-24 Machi 2007. Retrieved from http://www.tnrf.org/ files/S-INFO_Ole-Mwarabu.AK_2007_Taarifa_ya_uchunguzi_wa_uhamishaji_wa_wafugaji_toka_Mbarali_ Mbeya_kwenda_mikoa_ya_Lindi_na_Pwani_0.pdf.

Naguran, R. (2002). Property Rights and Protected Areas: The case of Ndumo Game Reserve. Paper presented at the research seminar on *Property Rights and Environmental Degradation*, organised by the Beijer International Institute of Ecological Economics 27-30 May 2002, Durban, South Africa. Retrieved from http://users.ictp. it/~eee/files/Naguran.pdf.

Nasi, R., Brown, D., Wilkie, D., Bennett, E., Tutin, C., Van Tol, G., & Christophersen, T. (2008). *Conservation and use of wildlife-based resources: the bushmeat crisis*. Secretariat of the Convention on Biological Diversity, Montreal, and Center for International Forestry Research (CIFOR), Bogor. Technical Series no. 33, 50 pp.

Natcher, D.C. & Hickey, C.G. (2002). Putting the Community Back Into Community-Based Resource Management: A Criteria and Indicators Approach to Sustainability. *Human Organization* 61(4):350-363.

Natcher, D.C., Davis, S., & Hickey, C.G. (2005). Co-management: Managing relationships, not resources. *Human Organization* 64(3): 240-250.

Ndetei, R. (2006). The role of wetlands in lake ecological functions and sustainable livelihoods in lake environment: A case study on cross border Lake Jipe - Kenya/Tanzania. In Odada, E. & Olago, DO. (Ed.) *Proceedings of the 11th World Lakes Conference*: vol. 2, 2006. pp. 162-168. Retrieved from http://193.190.8.15/odin/bitstream/1834/1492/1/WLCK-162-168.pdf.

Nelsen, F. (2004). *The evolution and impacts of community-based ecotourism in Northern Tanzania*. International Institute of Environment and Development. Working paper 131. Retrieved from http://www.iied.org/pubs/pdfs/9507IIED.pdf.

Nielsen, J.R., & Vedsmand, T. (1999). User participation and institutional change in fisheries management: a viable alternative to the failures of a top-down driven control? *Ocean and Coastal Management* 42(1): 19-37.

Nielsen, J.R., Degnbol, P., Viswanathan K.K., Ahmed, M., Hara, M., & Abdullah, N.M.R. (2004). Fisheries co-management – an institutional innovation? Lessons from south east Asia and Southern Africa. *Marine Policy* 28: 151-160.

Nilsson, M., & Segnestam, L. (2001). *Development and natural resources in the Mekong region: The institutional challenge*. World Resources Institute, 33. Retrieved from http://governance.wri.org/project_content_text.cfm?ContentID=2064.

Njaya, F.J., Donda, S.J., & Hara, M.M. (1999). *Fisheries co-management in Malawi: Lake Chiuta re-visit case study*. Paper presented at the International Workshop on Fisheries Co-management, Penang, Malaysia. December 31, 2001. Retrieved from http://www.comanagement.org/download/chiuta.pdf.

North, D. (1993). Toward a Theory of Institutional Change. In: Barnett, W., Hinich, M. and Schofield, N. (Eds.), *the political Economy Institutions, Competition, Representation*, Cambridge University Press.

Odera, J. (2004). *Lessons Learnt on Community Forest Management in Africa*. A report prepared for the project. Lessons Learnt on Sustainable Forest Management in Africa. Retrieved from http://www.afornet.org/images/pdfs/Community%20forest%20management.pdf. Accessed on 23 July 2008.

Olsson, P., Folke, C., and Berkes, F. (2004). Adaptive Co-management for Building Resilience in Social–Ecological Systems. *Environmental Management* 34(1): 75-90.

Ostrom, E. (1990). *Governing the commons: the evolution of institutions for collective action*. Cambridge University Press, New York. 280pp.

Ostrom, E. (1992). *Crafting institutions for self-governing irrigation systems*. San Francisco: Institute for Contemporary Studies.

Ostrom, E., Burger, J., Field, C.B., Norgaard, R.B., & Policansky, D. (1999). Revisiting the Commons: Local Lessons, Global Challenges. *Science* 284(5412): 278-282 Retrieved from www.sciencemag.org.

Owino, J. (1999). *Traditional and Central Management Systems of the Lake Victoria Fisheries in Kenya*. Socio-economics of the Lake Victoria Fisheries Project Report No.5. IUCN. Nairobi.

Peet, R., & Watts, M. (1996). Liberation ecology: development, sustainability, and environment in an age of market triumphalism. In: Peet, R., and Watts, M. (Eds.), *Liberation Ecologies: Environment, Development, Social Movements*. Routledge, London, pp. 1-45.

Pelesikoti, N. (2003). *Sustainable resource and environmental management in tonga: current situation, community perceptions and a proposed new policy framework*. PhD Thesis, Environmental science faculty of science university of Wollongong. 489 pp.

Peters, B.G. (1999). *Institutional Theory in Political Science*. The New Institutionalism. London: Pinter 183 pp.

Phillips, J.S. and Bhavnagri, N.P. (2002). The Maasai's education and empowerment. Challenges of migrant lifestyle. Childhood education. Spring 2002; 78, 3.

Pinkerton, E. (1987). Intercepting the State. Dramatic Processes in the Assertion of Local Co-management Rights. In: McCay, B.J. and Acheson, J.M. (Eds.), *The Question of the Commons: The Culture and Ecology of Communal Resources*. University of Arizona Press, Tucson, Arizona. pp. 344-369.

Pinkerton, E.W. (1994). Local fisheries co-management: A review of International Experience and their implications for Salmon Management in British Columbia. *Canadian Journal of Fisheries and. Aquatic Sciences* 51: 2363-2378.

Pinto da Silva, P. (2004). From common property to co-management: lesson from Brazil's first maritime extractive reserve. *Marine policy* 28 (2004) 419-428.

Plummer, R., & FitzGibbon, J.E. (2007). Connecting adaptive co-management, social learning and social capital through theory and practice. In: Armitage, D., Berkes, F., Doubleday, N. (Eds.), *Adaptive Co-Management: Collaboration, Learning and Multi-Level Governance*. University of British Columbia Press, Vancouver. 337 pp.

Pomeroy, R.S. (1995). Community-based and co-management institutions for sustainable coastal fisheries management in Southeast Asia. *Ocean and Coastal Management* 27(3):143-162.

Pomeroy, R.S. (1998). A Process for Community-based Fisheries Co-management. *Newsletter of the Asian Fisheries Social Science Research Network (AFSSRNews)*. Retrieved at http://www.seaturtle.org/PDF/Pomeroy_1998_AFSSRNews.pdf.

Pomeroy, R.S. (2001). Devolution and fisheries co-management. In: Meinzen-Dick, R. and Gregorio, M. (Eds), *Collective actions, property rights and devolution of natural resources management*. Proceedings, The international Conference, Puerto Azul, The Philippines, 21-25 June 1999, pp 111-146.

Pomeroy, L.R. & Albert, J.J. (1988). *Concepts of ecosystem ecology*. New York: Springer-Verlag.

Pomeroy, R.S., Sverdrup, J.S., & Raakjaer. N.J. (1995). *Fisheries co-management: a worldwide, collaborative research project*. Paper presented at Reinventing the Commons, the 5th annual conference of the Association for the Study of Common Property, Bodo, Norway, May 24-28, 1995, 21 pp.

Pomeroy, R.S. & Berkes, F. (1997). Two to tango: the role of government in Fisheries co-management. *Marine Policy* 21(5): 465-480.

Pomeroy, R.S., Katon, B.M., & Harkes, I. (2001) Conditions affecting the success of fisheries co-management: lessons from Asia. *Marine Policy* 25: 197-208.

Post, J., & Snel, M. (2003). The impact of decentralized forest management on charcoal production practices in Eastern Senegal. *Geoforum* 34: 85-98.

Pretty, J.N. & Shaxson, T.F. (1998). The potential of sustainable agriculture. *Newsletter of the Association for Better Land Husbandry* 8: 4-21.

Raik, D.B. (2002). *Capacity Building for Co-management of Wildlife in North America*. Cornell University. Human Dimensions Research Unit Series No. 02-2. 22 p.

Rie, O. (2002). Scrambling for land in Tanzania: formalization and legitimisation of land rights. *European Journal of Development Research* 14(2), 71-88.

Salum, A. (2007). *Sustainable wetland management in Tanzania. A case of Malagarasi-Muyovozi Ramsa Site (MMRS)*. MSc Thesis. Roskilde University Centre. 92pp.

Scarnecchia, D.L. (1990). Concepts of Carrying Capacity and Substitution Ratios: A Systems Viewpoint. *Journal of Range Management*, 43(6): 553-555.

Schusler, T., Decker, D., & Pfeffer, M. (2003). Social learning for collaborative natural resource management. *Society and Natural Resources* 15: 309-326.

Scott, J.C. (1998). *Seeing Like a State: How Certain Schemes to Improve the Human Condition Have Failed*. New Haven and London: Yale Agrarian Studies, Yale University Press, 1998. 445pp.

Sekhar, N.U. (1999). Decentralized Natural Re source Management: From State to Co-management in India *Journal of Environmental Planning and Management*, 43(1): 123-138.

Sekhar, N.U. (2004). Fisheries in Chilika lake: how community access and control impacts their management. *Journal of Environmental Management* 73: 257-266.

Sen, S., & Nielsen, J.R. (1996). Fisheries co-management: a comparative analysis. *Marine Policy* 20(5): 405-418.

Senyk, J. (2005). Lessons from the Equator Initiative: Community-based management by Pred Nai Community Forestry Group in the mangroves of Southeastern Thailand. Joint Project with the International Development Research Centre (IDRC) and the United Nations Development Programme (UNDP). Retrieved from www.umanitoba.ca/institutes/natural_resources/pdf/Tech%20Report%20Thailand%20-%20Jason%20Senyk.pdf.

Shackleton, S.E., Shackleton, C.M., Netshiluvhi, T.R., Geach, B.S., Ballance, A., & Fairbanks, D.H.K. (2002). Use Patterns and Value of Savannah Resources in Three Rural Villages in South Africa. *Economic Botany* 56: 130-146.

Shemdoe, R.S., & Mwanyoka, I.R. (2006). Traditional Knowledge and Practices in Managing Wetland Resources in Dryland Areas of Mpwapwa District, Tanzania. *Journal of African Affairs Institute of African Studies, Hankuk University of Foreign Studies, Korea.* 20: 181-200.

Sheridan, M.J. (2004). The environmental consequences of independence and socialism in North Pare, Tanzania, 1961-88. *Journal of History* 45, 81-102.

Shivji, I.G. (1997). Land tenure problems and reforms in Tanzania. In: Trux, A. (Ed.), *Land tenure issues in natural resource management in the Anglophone East Africa*, Paris: Sahara and Sahel Observatory (OSS), pp. 101-118.

Shukla, S. (2004). Strengthening Community-based Conservation through Traditional Ecological Knowledge. Natural Resources Institute, University of Manitoba, Canada. Retrieved from http://www.ed.psu.edu/ICIK/2004Proceedings/section9-shukla.pdf.

Sick, D. (2002). Managing environmental processes across boundaries: A review of literature on institutions and resource management. Minga programme initiative. International Development Research Council, Ottawa. Retrieved from http://network.idrc.ca/uploads/user-S/10378201220MEPfinal_Nov_-2002.pdf.

Singleton, S. (2000). Co-operation or capture? The paradox of co-management and community participation in natural resource management and environmental policy-making. *Environmental Politics* 9(2): 1-21.

Sneddon, C., Harris, L., Dimitrov, R., & Özesmi, U. (2002). Contested Waters: conflict, scale, and sustainability in Aquatic Socio-ecological system. *Society and Natural Resources* 15: 663-675.

Sokoni, C.H. (2001). *The Influence of Agricultural Marketing Reforms on Highland Farming Systems in Tanzania: The Case of the Uporoto Highlands, Mbeya Region*, Ph.D. dissertation. Dar es Salaam: University Press.

Sokoni, C.H. (2008). Commercialization of smallholder production in Tanzania: implications for sustainable resources management. *Geographical Journal* 174(2): 149-175.

Solecki, W.D. (2001). The role of global-to-local linkages in land use/land cover change in South Florida. *Ecological Economics* 37: 339-356.

Songorwa, A.N. (1999). Community-based Wildlife Management (CWM) in Tanzania: Are the Communities Interested. *World Development* 27: 2061-2079.

Sowman, M., Beaumont, J., Bergh, M., Baharaj, G., & Salo, K. (1998). An analysis of emerging co-management arrangements for the Olifants River harder fishery, South Africa. In A.K. Norman, J.R. Nielsen, & J.S. Sverdrup (Eds.), *Fisheries Co-Management in Africa*. Proceedings of a regional workshop on fisheries co-management research, Mangoshi, Malawi 18-23 March 1997. Fisheries Co-Management Research Project, Research Report No 12; pp 177-204.

Spielmann, R., & Unger, M. (2000). Towards a model of co-management of provincial parks in Ontario. *The Canadian Journal of Native Studies* XX(2): 455-486.

Spinage, C. (1998). Social change and conservation misrepresentation in Africa. Flora and Fauna International. *Oryx* 32(4): 265-276.

Spronk, S., & Webber, J.R. (2007). Struggles against accumulation by dispossession in Bolivia. The Political Economy of Natural Resource Contention. *Latin American Perspectives* 153, 34(2): 31-47.

Stratford, E., & Davidson, J. (2002). Capital assets and intellectual borderlands: socio-cultural challenges for natural resources management. *Journal of Environmental Management* 66: 429-440.

Sundet, G. (2006). Formalisation process in Tanzania. Is it empowering the poor? Retrieved from www.statkart.no/Tanzania+and+the+formalisation+process.d25-SwtnYXd.ips

Taiepa, T., Lyver, P., Horseley, P., Davis, J., Bragg, M., & Moller, H. (1997). Co-management of New Zealand's conservation estate by Maori and Pakeha: a review. *Environmental Conservation* 24(3): 236-250.

Tanzania Education Network (TEN) (2007). Strengthening education in Tanzania. CSO contribution to the education sector. Sector review 2006, Dar es Salaam.

Tanzania Electric Supply Company (TANESCO) (2000). The State of Environment in Pangani River Basin – A Reconnaissance Survey Report, October 2000.

Thompson, P.M., Sultana, P., Islam, N. (2003). Lessons from community based management of floodplain fisheries in Bangladesh. *Journal of Environmental Management* 69: 307-321.

Thompson, P.M., Sultana, P., Islam, M.N., Kabir, M.M., Hossain, M.M. & Kabir, M.S. (1999). *An assessment of co-management arrangements developed by the Community Based Fisheries Management Project in Bangladesh.* Paper presented at the international workshop on fisheries co-management, 23-28 August 1999, Penang, Malaysia. Retrieved from http://www.co-management.org/download/thompson.pdf.

Thomson, K., & Gray, T. (2009). From the community-based to co-management: Improvement or deterioration in fisheries governance in the Cherai Poyil fishery in the Cochin Estuary, Kerala, India. *Marine and Policy* 33 : 537-543.

Twong'o, T. K. and Sikoyo, G.M. (2001). Status of Lake Jipe ecosystem. An overview of the status of shared aquatic ecosystems in East Africa. The Plan of Action for the Development of Guidelines for Regional Environmental Impact Assessment of Shared Ecosystems of East Africa. Retrieved from http://www.icsf.net/icsf2006/uploads/resources/bibliography/docs/english/%3C1142405595426%3Eshared%20aquatic%20ecosystems%20east%20africa.pdf

Umar, B., & Kankiya, H. (2004). Beyond the Poverty Alleviation Programmes: Towards a New Framework for Managing Natural Resources in Nigeria. Deutscher Tropentag, October 5-7, 2004, Berlin, Rural Poverty Reduction through Research for Development. Retrieved from http://www.tropentag.de/2004/abstracts/full/100.pdf.

Uphoff, N. (1998). Community-based natural resource management: connecting micro and macro processes, and people with their environments. Plenary Presentation, International CBNRM Workshop, Washington, DC, 10-14 May 1998.

Western, D., & Wright, M. (1994). The background to community-based conservation. In D. Western, M. Wright, & S. Strum (Eds.), *Natural Connections. Perspectives in Community-based Conservation.* Washington, DC: Island Press, pp. 1-12.

Wiber, M.G. (1993). *Politics, Property and Law in the Philippines Uplands.* Waterloo Canada: Wilfrid Laurier University Press. 164 pp.

Willy, A.L. (2000c). The evolution of community based forest management in Tanzania. Retrieved from ftp://ftp.fao.org/docrep/fao/006/x7760b/X7760B18.pdf.

Wilson, D.C., Ahmed, M., Siar, S.V., & Kanagaratnam, U. (2006). Cross-scale linkages and adaptive management: Fisheries co-management in Asia. *Marine Policy* 30: 523-533.

World Commission on Environment and Development (WCED) (1987). *Our Common Future*, Oxford University Press. 43 pp.

Van der Knaap, M., Ntiba, M.J., & Cowx, I.G. (2002). Key elements of fisheries management on Lake Victoria. *Aquatic Ecosystem Health & Management* 5(3):245-254, 2002.

Yasmi, Y., Colfer, C.J.P., Yulian, L., Indriatmoko, Y., & Heri, V. (2007). Conflict management approaches under unclear boundaries of the commons: experiences from Danau Sentarum National Park, Indonesia. *International Forestry Review* 9(2): 597-609.

Yin, R.K. (1994). *Case Study Research: Design and Methods*, 2nd ed., Sage Publications, Inc., Thousand Oaks, CA.

Appendices

Appendix 1: Questions to farmers/livestock keepers/fishers at Lake Jipe

A: Personal Information

Name of household head:. . . . .

Name of the wife of the household head:. . . . .

Age of the household head: . . . .

Age of the wife of the household head: . . . .

Education of the household head: . . . .

Education of the wife of the household head:. . . . .

Occupation of the household head: . . . .

Occupation of the wife of household head:. . . . .

Tribe of the household head:. . . . .

Tribe of the wife of the household head: . . . .

Village: . . . .

B: Questions for farmers

1. What ways and methods do you use in farming?
   a. Flat cultivation
   b. Ridges
   c. Countour
   d. Crop rotation
   e. Intercropping
   f. Agroforestry
   g. Others (mention)
2. Why do you use these ways or methods?
   a. Improve soil fertility
   b. Control soil erosion
   c. Inherited from my parents
   d. Do not have alternative ways
   e. Other ways mention them
3. How big is your farm/what is the size of your farm?
4. How do you improve your farm?
   a. Apply manure (how often)
   b. Apply fertilizer (tell how much per season)
   c. Use pesticides (which, how often)
   d. Burn stalks
   e. I don't improve (why)
   f. Other ways/methods (mention them)

5. How do you prepare your farm?
   a. Slash and burn
   b. Mulching
   c. Set fires
   d. Make ridges
   e. Others (mention)
6. What imputs do you use in your farming?
   a. Hand hoe
   b. Tractor
   c. Plough
   d. Industrial fertilizers
   e. Insecticides
   f. Other (mention)
7. Where do you cultivate/practice farming?
   a. On the hills
   b. On the lowland
   c. Other places (mention them)
8. Why do you cultivate on the hills/sloping mountainous areas?
   a. Do not have farm plot elsewhere
   b. It is more fertile over there
   c. Easy to get water
   d. Other (mention)

C: Questions for livestock keepers

9. If you keep livestock, What ways or methods do you use in livestock keeping?
   a. Grazing
   b. Zero grazing
   c. Other (specify)
10. Why do you use these ways?
    a. Inherited
    b. Pasture shortage/unavailability
    c. Other reasons (specify)
11. What kind of livestock do you keep and how many are they for each category?
12. Are all these livestock yours or others belong to some other person?
    a. All mine
    b. Other person has entrusted his livestock to me
    c. Some are mine others belong to another person(s)
    d. Another answer (specify)
13. Where do you graze your livestock?
    a. On the mountains
    b. On the lowland
    c. Other areas (specify)

14. Why you graze your livestock in this area?
    a. Availability of pasture
    b. Water easily available
    c. Other reasons (specify)
15. What livestock keeping problems do you encounter?
    a. Livestock diseases
    b. Lack of water
    c. Low production
    d. Lack of markets for livestock and livestock products
    e. Fierce wildife
    f. Shortage of pastures
    g. Conflicts (mention them)
    h. Other problems (specify)
16. How do you solve the problems?
    a. Trap fierce wildlife
    b. Dig watering ponds
    c. Hunt fierce wildlife
    d. Plant pasture
    e. Purchase livestock medicines (veterinary medicines)
    f. Traditional veterinary (elaborate)
    g. Other techniques (specify)
17. If you keep livestock, what do you currently observe in livestock keeping?
    a. Shortage of water
    b. Declined production
    c. No changes
    d. Shortage of pastures
    e. Others (specify)
18. If you produce milk, how much do you produce?
    a. Less than 5 litres
    b. 5-10 litres
    c. 15-20 litres
    d. More than 20 litres
    e. Another amount

D: Questions for fishers

19. If you practice fishing, what ways do you use for fishing?
    a. Fishnets (what sizes)
    b. Traditional fish basket (Mgono)
    c. Hooks
    d. Other (mention)

20. Why do you use these ways/methods?
    a. To get abundant fish
    b. Inherited from my parents
    c. No alternative way
    d. Other (specify)
21. How many fish do you fish on average?
22. What other tools do you use apart from fishnets?
23. What problems do you face in fishing.
    a. Fierce wildlife
    b. Distortion of fishnets
    c. Limited fish (elaborate)
    d. Poor fish markets
    e. Others (specify)
24. How do you resolve these problems.
25. What changes do you notice in fisheries?
26. How is the availability of fish today when compared with past years?
    a. Increasing today (how) and why?
    b. Declining today (how) and why?
    c. No difference
    d. Other answer (specify).
27. What kinds of fish are available in the lake?

E: General questions for the three sectors.

28. How do your farming/livestock keeping/fisheries practices observe sustainable management of natural resources?
29. What seasons do you cultivate/fish/graze livestock on this area?
    a. Dry seasons
    b. Rain seasons
    c. Throughout the year
    d. Other times (mention it)
30. What problems do you face in farming?
    a. Insects pests
    b. Soil erosion
    c. Low production
    d. Lack of reliable markets
    e. Animal pests
    f. Others (mention)

31. How do you solve these problems?
    a. Traping the animal pests
    b. Use fertilizers
    c. Plant leguminous crops
    d. Hunt animal pests
    e. Other techniques (mention them)
32. Do you observe any changes in farming/fishing/livestock keeping?
33. What do you think are reasons for these changes?
34. How do you think negative changes, if any, can be alleviated or avoided?
35. How is the production in fisheries/livestock keeping/farming today as compared to the past years?
    a. Increased (mention scales)
    b. Decreased (substantiate)
    c. There is no difference
    d. Other answers (explain)
36. What conflicts do you encounter in your farming/livestock keeping/fisheries activities?
    a. Water conflicts
    b. Livestock destruction of crops
    c. Shortage of land
    d. Farm boundary encroachment
    e. Pollution of water
    f. Other (mention)
37. How is land tenure/ownership in your area?
    a. Traditional or custormary ownership
    b. Government ownership
    c. Other (mention)
38. What is the purpose of your farming/fishing/livestock keeping activities?
    a. Food
    b. Income/commercial
    c. Both
    d. Children school needs
    e. Other (mention)
39. How are the markets for your produce from farming/livestock keeping/fisheries activities?
    a. Good (how?)
    b. Poor (how?)
    c. Average (how?)
40. Where are the markets of your produce?
    a. Towns/urban areas
    b. In the village
    c. Other areas

41. How does the government support you in farming/livestock keeping/fisheries activities?
    a. Inputs
    b. Expertise
    c. Training
    d. Markets
    e. Others (mention)
42. When did you start farming/fishing/livestock keeping in this area?
    a. More than past 20 yrs
    b. Past 10-20 years
    c. 5 -10 years ago
    d. Less than 5 years ago
    e. Other (mention)
43. If you are a crops farmer, do you practice irrigation farming?
    a. Yes
    b. No
44. If you practice irrigation farming, where did you learn about it?
    a. Inherited from my parents
    b. From technical people (mention them)
    c. Others (mention)
45. Is the food you produce in the present years sufficient, increasing or decreasing? Please support
    your answer by giving esimations in comparison with the past years.
    a. Sufficient
    b. Increasing
    c. Decreasing
46. How much do you produce today?
    a. 5 -10 bags per acre (bags of what size)
    b. 10-20 bags per acre (of what size)
    c. More than 20 bags per acre
    d. Other amount (mention)
47. If you practice crops farming, what crops do you cultivate?
48. What other practices or activities do you do?
    a. Livestock keeping
    b. Fishing
    c. Technician
    d. Farming
    e. Other (mention)
49. If you practice fishing, how much fish do you catch today, and how do you compare the present
    catch with past catches? (please give estimations of catches based on your experience).
50. If you keep livestock, how is the status of production and productivity of the livestock? Is it
    increasing or decreasing?

51. In your opinions what should be done to improve farming/livestock keeping/fishing?
    a. Micro-credits to farmers/livestock keepers/fishers
    b. Training farmers/livestock keepers/fishers (on what)
    c. Improvement of inputs availability (what inputs)
    d. Other (mention them)
52. What are the realtionships between you and other resource users (e.g. farmers, livestock keepers and fishers)?
53. Are there any conflicts among users (livestock keepers, farmers and fishers)?
    a. Livestock browsing on crops
    b. Water pollution by the livestock
    c. Competing for water
    d. Competing for fishing space?
    e. Other (mention)
54. How do you resolve these conflicts?
    a. No solution
    b. Involve the village government
    c. Involve elders committees
    d. Press charges in legal courts
    e. Other ways (mention them)
55. What actors are involved in the resolution of the conflicts within, among and between sectoral resource users?
    a. Village government
    b. District government, farmers, livestock keepers and village government
    c. Farmers, livestock keepers and fishers
    d. Other actors (mention them)
56. How do these conflicts impact on the sustainable management of lake Jipe?
    a. Lack of participatory strategy for the management of the lake
    b. No impact
    c. Other impacts (mention them)
57. What are your opinions as regards the ways to do away with these conflicts?
58. Have the conflict brought only negative influences or they have lead to positive influences as well?
    a. Positive influences only (mention them)
    b. Negative and positive influences (mention them)
    c. Negative influences only (mention them)
59. How does the government assist you in the resolution of these problems and conflicts?
60. What traditional/customary ways are used in the management of natural resources and the resolution of the conflicts among the resources users at Jipe?
    a. Rituals (elaborate)
    b. Traditional rules (elaborate them)
    c. Avoid water pollution (how)
    d. Other regulations and rules

61. When did these arrangements and regulations start, and are they useful, and what benefits do you realize from the use of these traditional arrangements?
62. What are the strength and weaknesses of the rational institutions?
63. How are they in the present years in comparison with the past years?
64. What about their powers in comparison with the government regulations?
    a. Decreased
    b. Inceresed
    c. Other answer (elaborate)
65. Do you think the traditional regulations are still required today? And why you think so?
66. Do you think there are any land related problems in your fields or this area? If yes what problems?
    a. Soil erosion
    b. Declined fertility
    c. Shortage of land
    d. Other (mention)
67. What is the cause of these problems?
68. What do you think are the impacts of these problems?
69. What can be done to avoid these problems?
70. What problems are there as regards environmental degradation at lake Jipe?
    a. Siltation
    b. Water drying/drought
    c. Water weeds
    d. Decline of fishery
    e. Other (mention)
71. What are the causes of these problems?
    a. Poor farming practices
    b. Illicit or poor fishing practices
    c. Poor livestock keeping/grazing practices
    d. Drought
    e. Other (mention)
72. What impacts do the environmental problems have on the human beings and water and biodiversity and how?
73. You think these problems can be avoided? If yes, how? And if no, how?

How do you collaborate with the government to solve or alleviate environmental problems at Jipe area? How does the government support your efforts?

Appendix 2: Questions for focus groups discussions with various committees at Lake Jipe area

1. What is the name of your committee?
2. When did this committee start and what were the reasons for starting it?
3. How is the membership of this committee and how are the members appointed?
4. What are the roles and responsibilities of this committee?
5. What challenges, constraints and problems do you face in the implementation of your jurisdictions as a committee?
6. How do you resolve the challenges, problems and constraints you experience?
7. How do you link and cooperate with the government and users of natural resources?
8. What do you think should be done to improve the current situation?
9. How is the natural resources situation in this area? Is there any change in natural resources situation in the present days compared with the past years?
10. If you notice any change, which is it and what are the reasons for the emergence of this change?
11. What you think can be done to manage the change in the productive way for sustainability of the natural resources in the area?
12. How did you learn how to implement your responsibilities and activities? Where did you learn?
13. What do you think the government has to do in order to improve the situation of natural resources in this area?
14. Are there any other institutions you collaborate with in managing and enforcing sound natural resources management practices? If yes which are they, and how effective do you think this collaboration is?

Appendix 3a: Interview questions to the Ministry of Agriculture Food Security and Cooperatives, and Ministry of Livestock Development and Fisheries in Tanzania

1. What programmes/policies/strategies related to livestock keeping/agriculture/fisheries/water conservation do you have in your ministry?
2. When were these programmes/policies/strategies started, by whom and for what purposes?
3. Before these programmes/policies/strategies what other strategies/programmes and policies existed and why did you abandon or modify them?
4. What actors participated or participate in the formulation/modification of these programmes, and why involve these actors?
5. What achievements do you observe in the implementation of these arrangements/policies/ strategies/laws?
6. What constraints and problems do you face in the establishment, enforcement, and implementation of these arrangements or programmes?
7. What constraints, problems, assistance, and support do these policies, programmes, strategies of your sectors face or receive from other sectors? What are those that relate your sector with other sectors of water, fisheries, agriculture and livestock?
8. How does your plans/strategies/policies etc. help in conserving natural resources such as water, land? do you have any provisions for conserving other natural resources along the implementation of your sectoral activities? If not why, and if yes what are they about?
9. Do you think these plans/strategies/laws etc. need modification or revision? if yes, what and why?
10. What are the contradicting areas of your sectoral programmes and strategies as compared to those of other sectors? what are the effects or impacts of these contradictions in the implementation of your arrangements and programmes?
11. How do the international programmes and arrangements impact on formulation and implementation of your arrangements, strategies, and plans?
12. How did you involve the local governments in the formulation of your plans and strategies and programmes.

Appendix 3b: Interview questions for the fisheries, livestock and agricultural departments at Mwanga District in Tanzania

1. What programmes, policies and strategies related to livestock keeping/fisheries/water/agriculture do you get from the central government and which you are required to implement at the local level?
2. What programmes related with question (1) above do you formulate on your owns at the level of the local government?
3. How does the central government involve you in the formulation of the various sectoral strategies and policies and laws?
4. What support and assistances do you receive from the central government for the implementation of these policies, strategies and laws from the central government?
5. What are the institutions, mechanisms and laws do you formulate or design on your own to enable you in the implementation of the programmes and policies from the central government? How do these laws and mechanisms help you in the conservation of water and other natural resources and environment at lake Jipe?
6. Among the strategies and the programmes you have mentioned which are implemented at lake Jipe and along lake Jipe? how so these plans, strategies and arrangements relate with conservation od natural resources at lake Jipe area?
7. Since environmental and water conservation at lake Jipe depends on intersectoral collaborations, are there any intersectoral arrangements among these sectors at your level? What do they do? If they do not exist can you tell me why you do not have such arrangements?
8. What problems do you face or experience in the implementation of the arrangements from the central government?
9. Are there any conflicts, misunderstanding or contradictions and tensions among various sectors at your local government levels in the implementation of your sectoral programmes and policies and strategies? Please mention them and tell me what the sources of these tensions are?
10. What do you do in the resolution of these conflicts? Are there any participatory/cross-sectoral water and environmental management arrangements and of resolving the emerging tensions and conflicts among different sectors?
11. How do the various departments at the local level deal with or address the issue of environmental management at lake Jipe?
12. How do these tensions impact on sustainable environmental management at lake Jipe?
13. How do the various departments at the local government level collaborate to ensure that conflicts emerging in the management of lake Jipe are resolved?
14. How do you interact and communicate with community at lake Jipe in the implementation and enforcement of the arrangements related with natural resources management (fisheries management, livestock management and agricultural management). What approaches do you employ?
15. What problems and constraints do you encounter in your communication and interactions with the local community at Lake Jipe? and how do you solve these problems?

16. What issues are outside your rich you think if improved can improve the effectiveness of communication and enforcement as well as implementation of your plans, strategies and institutions within the community along Lake Jipe?

Appendix 3c: Interview questions for the village and ward government levels at Jipe

1. What are the strategies/plans/institutions regarding agriculture/livestock/fisheries come to the village community in this village?
2. How do the district departments involve you in the understanding and implementation of these programmes and plans in your villages?
3. How does the departments at the district level support and assist you in the implementation of these arrangements and plans in your communities?
4. What arrangements and laws do you make on your own so that to improve the effectiveness of the implementation of these plans and strategies and laws?
5. Who are involved in the formulation of these mechanisms or arrangements and why?
6. What are your opinions as regards the participation, compliance and response of the community in the implementation of these arrangements and plans?
7. If there is no good response or participation what do you think are the reasons? What you think has to be done so that to improve the community response?
8. Do you think the above mentioned plans and programmes are useful in improving socio-economic conditions of the people and in conserving the environment at Lake Jipe?
9. Who participate actively in the enforcement and implementation of these arrangements and plans and why you think is so?
10. What problems do you face in the introduction and implementation of these plans and strategies in the midist of the community? And how do you resolve these issues and problems?

Appendix 4: The list of the interviewed key informants at the Local government and central government level in Tanzania

Name	Department/Organizations	Designation
Mr. Gabriel Mramboa	Mwanga Natural Resources Department	District Natural Resources Office
Mr. James Tarimo	Natural Resources Department	District Fisheries Officer
Mr. Abdallah Munga	Agricultural and Livestock Departement	Ag District Agricultural and District Livestock Officer
Antony Dadu	Ministry of Livestock Development and Fisheries	Fisheries Officer
	Ministry of Livestock Development and and Fisheries	Policy planning officers
Mr G.R. Lyatuu	Jipe Ward	Jipe Ward Agricultural Extension Officer (WAEO)
Mr. A. Awadhi	Jipe	Village Agricultural Extension Officer
	Jipe village	(VAEO)
Mr.G. Madundo	MIFIPRO Trust Fund	The MIFIPRO Coordinator
Mr. R. Kidundi	Jipe Ward	Ward Executive Officer (WEO)
T.M. Kidaya	Jipe Ward	Ward Councillor
Mrs F. Wignessy	Jipe village	Village Executive Officer (VEO)
Mrs I. Mkamba	Kilimanjaro Region	Regional Agricultural Advisor (RAA)
Mr. M.I. Mziray	Environmental committee (Jipe)	Chairperson
A. Said	Butu village	Village Executive Officer (VEO)

Summary

Inadequate performance of natural resource management systems under the state monopolization has led to the evolution of collaborative natural resources management regimes in Tanzania. In past, since this country got independence in 1961 to late 1980s, natural resource management was under conventional state monopolized management regimes. Under these management systems, the community was regarded as a threat to the attainment of sustainable natural resource management. The government acted as a police ensuring that the community followed regulations, orders and institutional arrangements that were instigated by governmental authoritative agencies. The position of the community in these management regimes, therefore, was to obey the orders and directions from the government and when they did not comply they were eligible for punishment. However, because of its deficiency in terms of human and financial resources, the government could not appropriately and effectively control, monitor, and surveillance the vast natural resources systems whose degradation enhanced over time.

Continued degradation of natural resource under the state control, led to the change in governmental/national institutional arrangements and the introduction of regimes that involve the community adjacent to natural resources systems in their management. Lake Jipe is one of the natural resources in Tanzania in which the top-down management legacy dominated at post independent era. As other natural resources in Tanzania the deterioration of this lake has forced the government to change its approach and to accommodate the participatory natural resources management regimes. It is hypothesized that through the involvement of the local people in the proximity of these natural resources, the degraded resources can be restored, revitalized and sustainably managed.

Although various natural resource management institutional arrangements (e.g. legislation, policies, and Acts) provide for the involvement of the community in natural resources management, no research knowledge exists on how these collaborations occur and influence management of natural resource at lake Jipe social-ecological system in particular. This thesis investigated the possibility of the government and the local community for co-management of Lake Jipe social-ecological system in Tanzania. The main objective of this thesis was to determine collaboration between government and the community in the management of Lake Jipe social-ecological systems. Specific objectives entailed the determination of the collaborations between multiple governmental and community entities in the co-management of the multiple natural resources systems at Lake Jipe social-ecological system.

Social-ecological system, in the context of this study, is the system that comprises geographically connected upstream and downstream areas and which has multiple natural resources system. Contrary to conventional perspective of social-ecological studies in the co-management literature, which regarded social-ecological system as one natural resource system or a monolithic system, in this thesis social-ecological system is viewed as entailing multiple natural resources systems that though are different are interrelated.

In order to properly understand the complexities and processes of co-management of these multiple natural resource systems at Jipe social-ecological system, three concepts or vocabularies were used in this thesis. These concepts are co-management arrangements, institutions and actors. By co-management arrangements it implies sectoral natural resource management systems comprising community and governmental actors and institutions. The three arrangements which are a focus of this study are fisheries management system, use of natural resources in agriculture, and use of natural resources in livestock production. In these three co-management arrangements, collaborations among and between state/formal and community/informal as well as the bridging/co-management institutions were investigated on how they interact and govern the practices and decision making of community and governmental actors.

The methodological approach employed in the current study entailed the use of the geographic approach whereby upstream and downstream areas connected by flowing water and land resource were integrated in the investigation and multiple methods of data collection (triangulation) were used in the collection of data from multiple natural resources systems in these geographic localities. This approach enabled the use of multiple methods entailing observation, focus group discussion, household survey, participant observations and interviews with key informants. Lake Jipe in this thesis is viewed a case, from whose analysis generalization can be made for co-management arrangements of other social-ecological systems in Tanzania or elsewhere in the world.

The three co-management arrangements have depicted something different. In fisheries co-management arrangement, we see the existence of conflict between the resident fishers and non-resident fishers in their competition for spatial positions which are conceivably believed to have easy accessibility to the fish compared to other spatial positions. We have also found existence of conflicts between fisheries co-management institution and promoters of unsustainable fisheries practices (fishers) on the one hand and with governmental leaders on the other hand who seem to form alliances with illicit fishers in order to obtain illicit gains from these illicit fishers. The research has also uncovered the evolution and prevalence of cross-scale linkages between informal institutions (e.g. elders) and formal institutions (governmental regulations and actors) to combat illicit fishing practices. Also such cross-scale linkages have emerged entailing fisheries co-management institution (fisheries committee) and district fisheries authorities and administrators for combating unsustainable fisheries practices and mediating conflicts among resource users and between resource users and enforcers of the regulations. Generally it is observed in fisheries management that co-management initiatives have evolved because of complexities of problems in fisheries management and enforcement of sustainable fishing practices that could not be resolved by one institutional actor (i.e. the government alone or the community alone).

In the agriculture co-management arrangement the research uncovered the existence of complexities of issues and collaboration entailing governmental, non-governmental and community institutions and actors. Some of collaboration and linkages focus on specific resources whereas others focused on multiple natural resources in this co-management arrangement. Some co-management institution have evolved because some actors and institutions within specific boundaries of their operationing have not been able to address specific social conflictive problem that has traversed their capacity and therefore have formed alliance (co-management) with other institutions and actors in order to contain these challenges. This is the case in conflict resolution process for conflicts that occur in land under customary tenure management system. While the customary institution could not address these issues on its own, the government likewise could not do the same, and therefore efforts and capacities of the government and community (customary conflict managers) integrated into one co-management institution to assist each other to contain these challenges. In this arrangement also exist some informal ways for managing competitive resources (e.g. water) and mediating social relations among the resource users that links resources and social actors in geographically separated areas and in this the way trigger governmental actors to collaborate in the community initiated interactions. Besides, in this arrangement, there is an actor (MIFIPRO NGO) that enables and forges horizontal scale linkages among the community actors at two geographic areas (upstream and downstream) by improving their resource management practices, and at the vertical scale linkages between the governmental actors and community actors. This actor at the vertical scale enables effective operationing of government-community co-management institution (the environmental committee) by building the capacity of this institution through training awareness to enhance an understanding of actors over the governing rules they are supposed to oversee.

In livestock co-management arrangement we see the existence of heterogeneous ethnic actors depicting spatial and temporal co-existence and independent existence in various positions of the social-ecological system. Harmonious co-existence of these different ethnic groups at one spatial position is governmental-facilitated, and occurs when one spatial locality is deficient of pasture and therefore enforcing some ethnic actors to shift to other spatial positions where they encounter other settled actors. The two ethnic actors are the *Maasai* and the *Pare* tribes who exist at various spatial positions of the social-ecological system at some time when pasture is available on these different spatial positions but meet at some temporal space because as a result of pasture shortage the *Maasai* exercise nomadic livestock herding. Although in essence the two ethnic groups repulse each other, the intervention of authoritative governmental institutions facilitates their co-existence at the same temporal space of the year.

At the social-ecological system level, the research has identified institutions that play multiple positions and roles. Norms overseen by the elders in the Pare society for example mediate conflicts across the three co-management arrangements. Yet some institutions are specific in that they perform certain specific roles within sectoral boundaries. Fisheries co-management institution for example applies for fisheries co-management arrangement alone while reciprocity applies for livestock co-management arrangement alone likewise. In the integrated management of the social-ecological system the institutions that play cross-scale roles are important and therefore would need to be supported in order to enhance these important roles.

Overall the thesis concludes that there is no one co-management arrangement in Jipe social-ecological system but multiple co-management arrangements, at least three in the present thesis. The Jipe social-ecological system therefore cannot be understood properly when monolithic analytical approach is adopted because such an approach conceals inter and cross-scale interactions and diversities that exist in the system comprehension of which need disaggregating the system into diverse and multiple co-management arrangements.

It is recommended that further study is required to replicate the studied co-management arrangements and use of concepts of arrangements, institutions, and actors (elsewhere) in order to concretely understand institutional and ecological complexities that exist among these co-management arrangements and how they affect social-ecological system. It is further recommended that besides replicating this study in other different environments, the arrangements scope may be widened so that to analyze more than the three co-management arrangements and their influence on co-management of social-ecological system. Besides, further study is recommended on the effectiveness of institutions in order to under which or the institutions have positive impacts and which have negative impacts. This will enable us to capitalize the positive impacts and to strategize and plan on turning negative impacts to positive influences so that to improve sustainable management of social-ecological systems in Tanzania and elsewhere around the global.

About the author

Christopher Paul Mahonge was born in 9th of February 1970 in Tanzania. He did his primary education in Mgolole primary school in Morogoro region in Tanzania. Afterwards he pursued his secondary (1985-1988) and high school (1989-1991) education at Mzumbe Secondary School in Morogoro region in Tanzania. He then joined Sokoine University of Agriculture for Bachelor of Science in Forestry (1992-1995) and Master of Science in Forestry (1998-2001). After MSc studies (2002-2003) the author was employed by Sokoine University of Agriculture in Tanzania as an Assistant Researcher whereby he worked in a project on sustainable rural development through a collaboration of Sokoine University of Agriculture, Japan International Cooperation Agency (JICA) and Kyoto University of Japan. From 2003 April to-date the author works as Research Fellow with Sokoine University of Agriculture. In September 2005 the author joined Wageningen University for PhD studies.